The Biology of Protozoa

Michael A. Sleigh

Ph.D., D.Sc.

Professor of Biology, University of Southampton

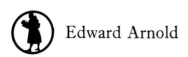 Edward Arnold

First published 1973
by Edward Arnold (Publishers) Limited
41 Bedford Square, London WC1B 3DQ

Reprinted 1975, 1979
Reprinted with additions 1981

Boards Edition ISBN: 0 7131 2409 1
Paper Edition   ISBN: 0 7131 2410 5

Printed in Great Britain by
Whitstable Litho Ltd
Whitstable, Kent

Preface

Research using Protozoa has flourished in recent years not only because of studies directed at increasing our knowledge of the Protozoa themselves, but also because many people have recognized that these organisms provide excellent subjects for studies of general biological phenomena at the cellular level—the possibility of obtaining large numbers of identical cells, uncontaminated, and even at the same stage in their growth and division cycle is particularly valuable in biochemical studies; studies on Protozoa have therefore made a substantial contribution to modern Cell Biology. This flowering of research activity has principally been directed towards elucidating the structure and understanding the functioning of Protozoa as cells—their organization, biochemistry and physiology, but new research in such fields as ecology and genetics has also revealed more about the importance of Protozoa in the economy of nature and about the variety of methods evolved by living things to solve problems of living. We are therefore at an exciting stage in the development of the study of Protozoa.

The growth of Protozoology as a substantial discipline has been linked with the appearance of several excellent textbooks; in recent years the two volumes edited by Grassé,[90] and books by Grell,[92] Mackinnon and Hawes,[165] Sandon,[222] Dogiel[67] and Kudo[145] have given coverage to various aspects of the study of Protozoa, but only bring the reader to the threshold of modern studies in Cell Biology. Several recent texts are valuable but are more limited in scope, covering ultrastructure in the book by Pitelka,[198] only parasitic species in the book by Baker[9] and only a few species in the book by Vickerman and Cox.[274] Accounts of recent research are therefore not generally available in textbooks, and one has to depend upon specialist reviews and original papers to provide a reasonable coverage of the subject. Comparison of the recent review articles in the four volumes of *Research in Protozoology*[32] with their predecessor *Protozoa in Biological Research*[30] shows the enormous development of many branches of the subject, but all of these articles and those in the three volumes of *Biochemistry and Physiology of Protozoa*[118a, 118b, 162a] have provided much information for this book. The increase in research activity has led to an enormous literature about Protozoa, so that the references quoted have been restricted to major works, reviews and other recent papers that refer to earlier studies and provide an introduction to literature that could not be listed in the bibliography.

It is the main aim of this book to emphasize new insights into cell functioning and ultrastructure as they apply to Protozoa, and to illustrate the wealth of diversity in organization and physiology that has been revealed among the various groups of organisms classed as Protozoa. The background of systematic Protozoology is related with modern developments in the hope that this book will provide students with a guide to many aspects of new research on Protozoa.

It is a pleasure to acknowledge the help given by many people in the preparation of this book. I owe a particular debt of gratitude to Mrs. Margaret Attwood whose skilful drawings illustrate the diversity of flagellates, amoebae and ciliates in Chapters 5, 7 and 8; almost every drawing incorporates information taken from several sources, so that in few cases is it possible to acknowledge a single origin for the figure. I should also like to thank the following who generously gave me copies of electron micrographs for illustrations that appear in the book: Drs. L. H. Bannister, C. F. Bardele, J. and M. Cachon, R. M. Crawford, J. D. Dodge, A. E. and A. G. H. Dorey, R. Gliddon, H. Jørgen Hansen, R. H. Hedley, A. Hollande, B. S. C. Leadbeater, A. C. Macdonald, J. W. Murray, B. Nisbet, E. B. Small, J. B. Tucker and K. Vickerman. The permission from authors and publishers for the reproduction of figures previously published elsewhere is also acknowledged with gratitude. A number of secretaries have assisted at various stages, and I am particularly grateful to Miss Gaye McCrindle, whose accurate work, patience and attention to detail greatly aided the preparation of the final manuscript; Miss Jeanette Date and Miss Margaret David have also given valuable help. I wish to express my thanks to all those colleagues who have helped by answering questions, discussing evidence and reading sections of this book; their comments have added to my understanding and have contributed substantially to the accuracy of many statements. The wise guidance of Professor E. J. W. Barrington, the General Editor, is gratefully acknowledged.

Bristol 1973   M.A.S.

Table of Contents

I

An Introduction to Protozoa

The name Protozoa, which means 'first animals', is an appropriate collective name because many members of the group have the simplest organization of any animals, and as far as we know the first animals to have appeared in evolution would come within this taxonomic category. The Protozoa themselves have evolved: there is now a considerable diversity of organisms within the group, and all multicellular animals are believed to have originated from Protozoa. The group comprises four quite distinct classes of organisms, three of which, the Ciliophora, Sporozoa and Cnidospora, are entirely animal, while the Sarcomastigophora contains both animals and plants. Many taxonomists feel that the organization of these four classes is sufficiently different for each of them to be given the status of a phylum; however, in the majority of classification schemes they appear as sub-phyla of the phylum Protozoa, and this convention will be followed here.[39]

FEATURES OF THE FOUR SUB-PHYLA

The groups of photosynthetic flagellates, incorporating those colourless flagellates which have close relationships with these groups, are conventionally placed in zoological classifications in the Phytamastigophorea (plant flagellates).[115] Other colourless flagellate Protozoa which do not have features indicating a close relationship with any algal group are placed in the Zoomastigophorea. It is likely that this assemblage as at present constituted is polyphyletic and that several, if not all, sections of it will ultimately be classified in the same series as the groups of algal flagellates. Several of these groups of animal flagellates include specialized symbionts, many of them parasites in animals, and in some groups the organisms carry large numbers of flagella, while the algal flagellates commonly have two and seldom more than four flagella.

A diagram of the structure of a simple chrysomonad flagellate is shown

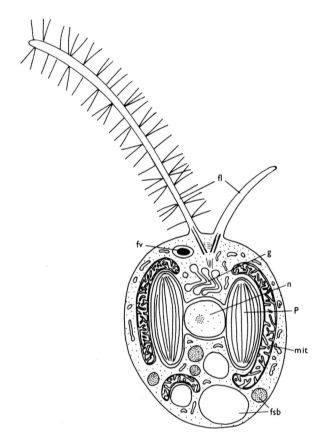

Fig. 1.1 Diagram of structures found in *Ochromonas*, a flagellate member of the Chrysophyceae, showing the flagella fl, food vacuole fv, golgi body g, nucleus n, plastid p, mitochondrion mit, and food storage bodies fsb; length of cell body about 10 μm.

in Fig. 1.1. The cell possesses the basic components of a phytoflagellate, including flagella, nucleus, plastid, food storage bodies, mitochondria, vacuoles, vesicles and other membranous and fibrous inclusions. Such structures tend to be found in characteristic positions within the cell of a particular species of flagellate. The cell surface in this case is not structurally specialized.

The amoeboid Protozoa (Sarcodina) are now placed close to the flagellate organisms within the sub-phylum Sarcomastigophora,[115] for many organisms show both flagella (on non-gametic cells) and amoeboid pseudopodia, together or at different phases of the life cycle. It is likely that several parallel lines of evolution have led from pigmented flagellates to the various groups of amoeboid Protozoa. There are photosynthetic amoeboid forms

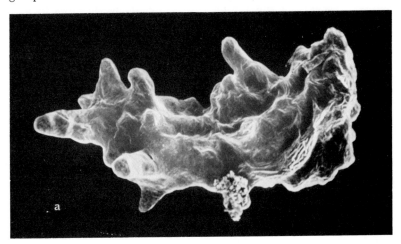

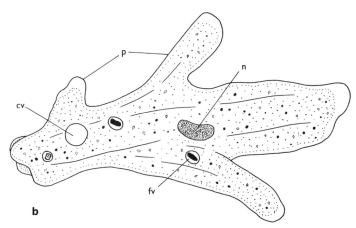

Fig. 1.2 *Amoeba proteus*. (a) Scanning electron micrograph (kindly supplied by E. B. Small and D. S. Marszalek).[242] (b) Diagram showing pseudopodia p, contractile vacuole cv, nucleus n and food vacuole fv; cell length about 500 μm.

in the Chrysophyceae, but generally amoebae are phagotrophs and clearly animals. The primary characteristic of sarcodine Protozoa is the possession of some form of pseudopodium, but there is considerable diversity of structure, notably in the character of any test, shell or skeletal material that may be present, and in the type of pseudopodium, e.g. broadly lobed, needle-like or reticulate. The diagram of *Amoeba proteus* in Fig. 1.2 shows how little structural organization an amoeba may have in comparison with a flagellate or ciliate protozoon. The cytoplasm is subdivided into the superficial ectoplasm and the internal endoplasm which flows forward through the body into the cylindrical pseudopodia. Within the endoplasm are found the nucleus, food storage granules, mitochondria, vacuoles and other membranous or crystalline inclusions.

Many taxonomists are now agreed that the opalinid Protozoa (Chapter 6) are more suitably classed in the Sarcomastigophora than with the Ciliophora;[37] they share with ciliates the possession of rows of cilia, but in nuclear and reproductive features they appear to be more like flagellates. Ecologically they form a very restricted group and no close relations with any other protozoan group can be recognized. Conceivably they may have evolved in parallel with the ciliates from a common flagellate ancestor.

The Sporozoa are a group of specialized parasitic Protozoa with complex life cycles, initially classed together because they showed a common mode of life and were not like any other types of single-celled organisms. A diversity in certain features led to the suggestion that the group is polyphyletic in origin. However, recent studies of fine structure have revealed the presence of certain unique organelles in members of several groups,[278] so perhaps the Sporozoa are really phylogenetically compact. They are haploid organisms with a zygotic meiosis (see Chapter 4), a feature known to be shared only with certain flagellates; many Sporozoa have flagellated gametes.

Members of the Cnidospora have distinctive spores whose infective contents are amoeboid sporoplasms, in contrast to the regular-shaped sporozoites of Sporozoa (Fig. 1.3), and whose elaborate structure is in many forms produced by differentiation of several cells derived from a multinucleate plasmodium. They are clearly distinct from the Sporozoa, and some authorities doubt that they are Protozoa at all because of their multicellular spores. All members of the sub-phylum are parasitic, occurring in cold-blooded vertebrates and in invertebrates. Origins as diverse as amoebae, dinoflagellates and several metazoan groups have been suggested for this group, but their specialized structure has so far revealed no likely relationships. The polar capsules of the Cnidospora are reminiscent of nematocysts, and it will be interesting if they are found to show ultrastructural details comparable with the nematocysts of coelenterates or dinoflagellates.

The ciliates (Ciliophora) clearly form a natural group isolated from other Protozoa, and indeed from other living organisms, by a number of specialized features. They have a pellicular organization based primitively on rows of cilia, and are unique in their nuclear dimorphism and in the sexual process of conjugation. It is likely that the ciliates evolved from a flagellate stock by multiplication of the locomotor structures, and that the nuclear dimorphism occurred later since some simpler ciliates have only one unspecialized type of nucleus. Ciliates are diploid with a gametic meiosis, a pattern shared among Protozoa with opalinids, heliozoans and some zooflagellates.

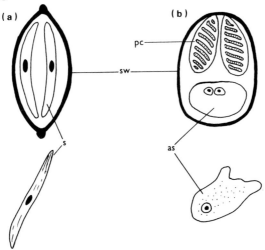

Fig. 1.3 Comparative diagrams of members of (a) Sporozoa and (b) Cnidospora, showing above characteristic spore structures (spore wall sw, sporozoite s, polar capsule pc, and amoeboid sporoplasm as) and below the appearance of infective cells that leave the spores; length of spores about 10–25 μm.

The structure of a representative ciliate shown diagrammatically in Fig. 1.4 illustrates some common features of ciliates. The body surface is covered with cilia which are mostly aligned in rows called kineties. This pellicular kinety pattern is interrupted in the region of the cell mouth (cytostome) where there may be specialized compound cilia which are used in feeding; the depression in which these compound cilia lie, and which leads into the cytostome, is called the buccal cavity. The pellicle is a specialized superficial zone of cytoplasm including the 'infraciliature' (ciliary bases and attached fibres) and often membranous structures, some mitochondria and other organelles. The interior cytoplasm is more fluid and mobile; it contains the macronucleus, micronucleus, food vacuoles,

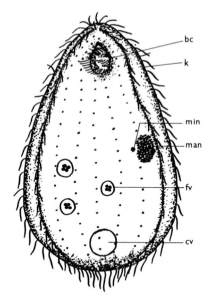

Fig. 1.4 Diagram of the ciliate *Tetrahymena* showing kinety rows of body cilia k (only positions marked in most rows), buccal cavity bc with feeding cilia, micronucleus min, macronucleus man, food vacuoles fv and contractile vacuole cv; cell length about 50 μm.

mitochondria, and many other inclusions. Contractile vacuoles are often closely associated with pellicular structures, although they protrude into the interior cytoplasm.

RELATIONSHIPS OF PROTOZOAN GROUPS

It is believed that the first cellular organisms were procaryotic, without membrane-enclosed organelles in the cytoplasm, and therefore lacking the true nuclei, mitochondria and plastids which are found in cells of eucaryote organisms. Moreover, the flagella found on bacteria (procaryotes) are simpler fibrous organelles than the complex flagella built of a bundle of 9+2 longitudinal fibrils (p. 28) which are found on eucaryotic cells; flagellar organelles of both types are locomotory structures, but it is unfortunate that organelles of quite different construction, and associated with organisms at two different evolutionary levels, should be called by the same name. Such procaryotic organisms as bacteria and blue-green algae (Cyanophyta) may possess photosynthetic pigments, which in Cyanophyta include chlorophyll *a*, and they may have internal membranes in simple formations, notably the photosynthetic lamellae of blue-green algae, but these lamellae are never grouped into membrane-bound plastids.

Genuine plastids, in which at least two unit membranes enclose the chlorophyll-bearing lamellae, show their simplest form with single, separate thylakoids (p. 105) in the red algae (Rhodophyta). These are eucaryotes, but lack the 9+2 flagella found in all the other major groups of eucaryotes; this and the form of their plastids suggest that the Rhodophyta is the most ancient eucaryote group.

The remaining algae (Contophora) have 9+2 flagella, at least at some stage in some members of each group, and have thylakoids adhering in groups within the plastids. A major distinction between two taxa of these algae is the appearance of chlorophyll *b* as one of the photosynthetic pigments in a few groups, collectively called the Chlorophyta, while the pigments of the Chromophyta do not include chlorophyll *b*. Other characteristics of these taxa and of their constituent groups will be considered later (Chapter 5). Included among the Chromophyta are several groups, notably the diatoms (Bacillariophyceae) and the brown algae (Phaeophyceae), which do not have a dominant flagellate stage and which are never included in the Protozoa. Most of the other groups are frequently classified with the Protozoa[115] as well as with the Algae,[216] and this is particularly true of those groups in which some members have lost their photosynthetic pigments and taken up saprophytic or holozoic nutrition; many dinoflagellates, for example, have such a preponderance of animal characteristics that they seldom appear in books concerned with Algae.

These early eucaryotic organisms stand at the level of the evolutionary bifurcations between animals and plants. The evolution of animals from plants has taken place many times among these groups. It is therefore necessary for the whole assemblage of algal groups to be included in a single phylogenetic scheme with Protozoa if the true relationships of many holozoic flagellates are to be understood. Many conventional classifications of Protozoa draw artificial boundaries between Algae and Protozoa within the groups of flagellated Algae, but in this book an attempt will be made at integration.

While many of the algal groups do not seem to be close to the lines of evolution leading to other groups, it is likely that all other eucaryotic organisms may have their origins somewhere among the Chromophyta or Chlorophyta. The land plants possess chlorophyll *b* and other features which link them closely to the Chlorophyta, and in particular the Chlorophyceae. The specialized features of such protozoan groups as the Ciliophora or the Sporozoa on one hand and of the pigmented flagellates on the other make it difficult to suggest probable relationships between these Protozoa and any of the algal groups. However, it seems likely that Algae of the Chromophyta, which is probably a more ancient group than the

Chlorophyta, are closer to the animals, and that among chromophyte Algae the Chrysophyceae are in several respects good candidates as possible ancestors of a variety of holozoic Protozoa. It is thought that ciliates, sporozoans and various groups of amoeboid and flagellate Protozoa may have evolved independently at different stages in the evolution of the chromophyte Algae. Indeed it is suspected that electron-microscope studies of some small groups of colourless flagellates (which at present

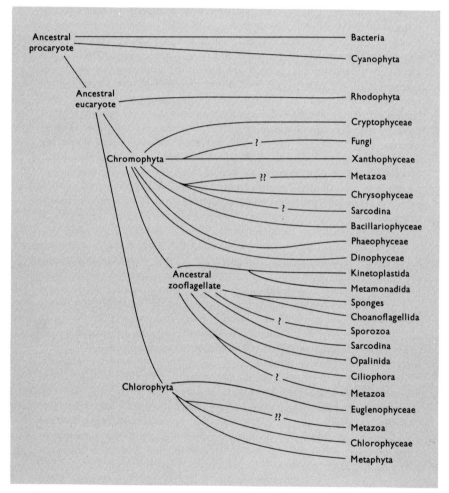

Fig. 1.5 Supposed phylogenetic relationships between various groups of organisms mentioned in the text.

occur almost exclusively in protozoan classifications) may soon show that they are more closely related to a group or groups of chromophyte algae than to other colourless 'zooflagellates'. It is quite likely that fungi are also related with the chromophyte algae in an evolutionary relationship parallel to that with Protozoa.

It is widely believed that multicellular animals evolved from protozoan organisms, but which group or groups of Protozoa may have been involved has been the subject of much controversy.[96, 133] Most agreement concerns the likely connection between the collar flagellate Protozoa (among the Zoomastigophorea) and the sponges, since both groups contain flagellate cells of the same unusual type. Sponges, however, are not generally considered to be ancestral to any of the metazoan phyla, so other possible routes of evolution should be considered. Some authorities support the idea that a colonial phytoflagellate could have been the evolutionary ancestor of the first coelenterates, perhaps through an intermediary planuloid form. Another theory is that the first metazoan organisms were the acoelan turbellarians, and that these appeared as a result of the cellularization of primitive, multinucleate, ciliated Protozoa. Yet other possibilities have been suggested, including the hypothesis of a polyphyletic origin of the lower metazoan phyla from several groups of Protozoa (see also [41a]).

Some of the supposed relationships between groups of lower organisms that have been discussed here are shown in Fig. 1.5. While it is reasonable to be confident about some of the broader generalizations concerning the place of Protozoa in the evolution of living organisms, many attempts to infer relationships from the characteristics of present-day forms must be speculative. The divergence between the major groups of lower organisms is extremely ancient; there are reports of fossil flagellates from deposits well over 2000 million years old,[166] but the remains of pre-Cambrian rocks are generally fragmentary. Most of the animal phyla had evolved by the beginning of the Cambrian era, nearly 600 million years ago. It is clear that evolution is continuing at a faster or slower rate in all groups of Protozoa, and it is likely that any organisms which would have provided valuable phylogenetic links may have been eliminated long ago by more successful species.

PROTOZOA AS CELLS

The form of organization of the protozoan body has been the subject of much semantic discussion.[119, 165, 198] The simplest organisms of all major groups, except perhaps the Cnidospora, are clearly unicellular; they show at the ultrastructural level many of the same characteristic features as cells of Metaphyta and Metazoa—the same membrane formations and organelles made from the same types of components. In several protozoan groups organisms have evolved extremely elaborate organelles of such a diversity of types within the cell membrane that these cells are the most complex known. Many protozoan cells are very large, and this increase in size apparently demands an increase in nuclear material in the form of duplicate sets of chromosomes, either within a polyploid nucleus or in multiple nuclei. The formation of colonies of protozoans is common in many groups, and probably originated by incomplete separation following division of the organism. Organisms of several groups show an approach to genuine multicellular organization, including a division of labour of partially differentiated cells, within the most integrated types of colony. Many biologists have argued that such multinucleate and multicellular Protozoa indicate that the members of this phylum are not unicellular, so that they prefer to refer to the Protozoa as acellular or non-cellular. However, it is clear that these more complex forms are no more than specialized products of the evolution of organisms whose organization is based upon the elaboration of the components of a single cell.

A protozoan cell is a complex, highly integrated, unit of life. Its functioning depends on a turnover of chemical materials, a store of biochemical information and a through-flow of energy. All three are normally necessary for maintenance as well as growth and other characteristic activities of the organism. Organic molecules which are incorporated into the structural elements and many other vital components of the cell may be built up within autotrophic organisms from such simple molecules as carbon dioxide, water and salts containing nitrogen, sulphur, phosphorus and other elements; the synthesis of organic compounds from such simple precursors occurs in association with the fixation of light energy in the plastids of photosynthetic cells, but similar synthesis may utilize energy derived from the breakdown of simple organic compounds (Chapter 3). Heterotrophic Protozoa take in large molecules both as a source of raw materials for cell construction and as a source of energy; usually their powers of synthesis of small organic molecules are limited. While the small molecules required by autotrophic organisms may enter the cell by diffusion, the uptake of complex molecules requires an active process and usually involves taking the food substance into vacuoles in the cell, and the subsequent enzymic breakdown of macromolecules into smaller units which may be passed through the vacuole membrane. The cell membrane forms the zone of contact between a protozoan cell and the environment; everything that enters or leaves the organism must pass through this membrane, so that it is a vital controlling region exerting an important influence on intracellular concentrations. Agents which act on the protozoan must either act at the cell membrane or must pass through the membrane to an intracellular site of action.

Energy derived from light or the breakdown of organic molecules is stored in the protozoan cell in the chemical bonds of organic compounds. Some of these compounds are labile and carry a relatively large amount of readily available energy; the best known example is adenosine triphosphate (ATP), in which the energy is carried in 'high energy' phosphate bonds. Many other compounds which serve as stores of energy are inert and the energy they contain is mobilized by the enzymic oxidation processes associated with respiration and is transferred to energy carriers like ATP. Aerobic respiration involves the utilization of oxygen and is generally associated with mitochondria in the cytoplasm. Energy given up by ATP and similar carrier molecules is used in protozoan cells for synthesis and for movement. It is likely that energy is normally used in the formation of chemical bonds: this is obvious in the case of molecular synthesis, and in movement it is thought that the changes in shape or position of molecules which lead to a diversity of movements are also the result of the formation of temporary intermolecular or intramolecular chemical bonds. Most reactions in protozoan cells involve the loss of some energy and the energy released on the breakage of many bonds may not be recoverable.

The information store necessary for molecular synthesis within Protozoa as within other cells is provided by 'blueprints' carried in the genes of the nucleus. These blueprints take the form of coded sequences of subunits in the molecules of DNA (deoxyribonucleic acid) that constitute the genes. The information encoded in DNA finds expression in the synthesis of proteins; during this synthesis the information passes through two stages of transcription in which reactions involving three types of RNA (ribonucleic acid) and many enzymes build up a polypeptide chain whose sequence of amino acids is specified by the code of the DNA. The final stage of this synthesis is associated with ribonucleoprotein bodies called ribosomes, the majority of which occur in the cytoplasm. Some of the proteins synthesized are structural proteins, but the majority are enzymes functioning in the promotion of a variety of cell reactions, including the synthesis of molecules of many types. Molecules synthesized in protozoan cells are mainly used within the cells, but most Protozoa produce some forms of extracellular products, including extracellular envelopes, cyst walls, extrusive organelles and mucous material.

The character of a protozoan cell depends on the types of proteins it contains. Individuals within a species may show minor variations in their total protein constitution; if the differences between the protein constitutions of two populations of a protozoan species become extensive, it is likely that the two populations will no longer interbreed, and therefore that they become separated as two distinct species. The differences between the protein constitutions of two species will be greater or less according to the remoteness or closeness of their phylogenetic relationship. Most protozoan cells change in character, because of changes in the pattern of protein synthesis, during their life cycle; e.g. there is usually a period in which shells, skeletons or new cell organelles are formed, in species of many groups there is a cyst stage, and in the case of Sporozoa the organism goes through a sequence of different morphological stages. A cell is believed to carry the same gene pattern in the nucleus throughout its life cycle, but to utilize different assortments of these genes at different times so that a different character will temporarily be imparted to the cell by the different proteins synthesized. A complex interaction of internal and environmental factors is believed to determine which genes are used for protein synthesis at any particular time.

The provision of a complete information store for each daughter cell at the division of a protozoon is made possible by the exact replication of the DNA molecules before division of the nucleus. This precise replication and the high degree of stability of the DNA molecule are vital for the inheritance of specific features and the conservation of the characteristics of a species. However, DNA molecules are not absolutely stable and mistakes in replication do occur. Variations (mutations) can occur within the gene pattern of the cell and can be inherited to provide variants in cell character which form the raw material of the process of natural selection. Variation in gene patterns is necessary for evolution to take place. Some Protozoa contain only a single set of chromosomes (haploid cells), and in these forms every gene may find expression and be 'tested' for its usefulness to the organism; other forms contain two or more sets of chromosomes (diploid or polyploid cells), and a mutant gene may or may not find expression depending on the nature of the homologous (allelic) gene(s) on the other chromosome(s). In haploid organisms only successful mutations survive, but unsuccessful genes may persist in diploid organisms. Many genes are only of value when they occur in suitable combinations with other genes; many enzymes are involved in chain reactions where each step is essential for completion of the sequence. The occurrence of several suitable mutations simultaneously, so as to make possible the establishment of a new reaction sequence, is a very unlikely event, and in haploid organisms demands very large populations or results in very slow evolution. The chance of the occurrence of suitable combinations of mutations may be increased in diploid or polyploid organisms, but the most important means of bringing about increased variation in the gene content of organisms are those processes associated with sexual fusion (Chapter 4). The assortment of genes in meiosis and the recombination of genes at fertilization provide great plasticity of the gene pattern within an interbreeding population of organisms. The evolution of sexually reproducing Protozoa can be much more rapid than that in species which only reproduce asexually.

In subsequent chapters this introduction to cellular organization of

Protozoa will be expanded by discussions of the structure and functions of the component organelles of protozoan cells, the nutrition and the reproduction of Protozoa. The characteristics and range of form of the various major groups of Protozoa will be described, and in the final chapter there is a discussion of the ecology and biological and economic importance of members of the group.

2

Features of Protozoan Organization

The major features of body structure in Protozoa are at the organelle level.[94, 198] Most of the cytoplasmic structures which are found in other eucaryote cells are also recognizable in Protozoa. These structures may be classified as membrane elements and fibrous elements, although the nuclear contents and extracellular products of cells do not entirely fall within these categories.

MEMBRANOUS STRUCTURES

Membrane elements in cells are composed of lipoprotein structures about 7·5 nm thick, which appear in electron micrographs of sectioned membranes as two dense lines each about 2 nm thick separated by a light space 3–4 nm thick.[213] These zones are believed to represent parts of a sandwich structure composed of a double layer of lipid molecules, orientated to provide a hydrophobic centre to the membrane, linked on either side at the hydrophilic ends of the lipid molecules to a layer of protein molecules. There is some recent evidence that the structure of at least some cell membranes is more complex than this. The protein material (sometimes with carbohydrate components) may be different on the two sides of the membrane, so as to impart an asymmetry to the structure. Such a 'unit membrane' structure has impermeability (provided by the lipid) and strength with elasticity (provided by the protein), two characteristic properties which are particularly necessary for the surface membrane of cells.

The cytoplasm of every protozoon is completely enclosed in a unit membrane layer, called the **plasma membrane**. Through this membrane must pass all substances which enter or leave the cell, so that the properties of the membrane determine the character of the cell by controlling the concentration of substances within it. The membrane is very permeable to some substances which may enter or leave readily, and is very impermeable to other substances, which may not be allowed to enter or may be kept within the cell; some substances to which the membrane has a low permeability may be actively transported across the membrane with the use of metabolic energy. Such differential permeability of the cell membrane makes it possible for the cell to accumulate certain substances so that a suitable chemical environment may be provided for intracellular processes; the semipermeability of the membrane often also necessitates some regulation of the osmotic concentration of the organism, particularly in fresh water. The plasma membrane forms a most important part of the homeostatic machinery of Protozoa, as it does also in cells of other types. The low permeability of the membrane to certain ions and the ability to transport ions actively have made it possible for many cells, including Protozoa, to maintain an electrical potential across the plasma membrane and some other membranes in cells, and to use these membrane potentials in a number of physiological activities.

It is also important that since the plasma membrane is the region of contact with the outside world, external agents which exert an effect on the organism do so at the membrane or after penetrating through the membrane. The membrane probably has little control over electromagnetic and other radiations, but a large proportion of other influences or substances probably 'distort' the membrane physically or chemically in order to achieve their characteristic effect.

While much of the information about the plasma membrane owes relatively little to the advent of the electron microscope as a tool for biological investigation, our knowledge of the internal membranes of cells is very largely the result of recent fine-structure studies; the general form of some of these membranous structures is shown in Fig. 2.1. The *endoplasmic reticulum*, whose existence was only suspected previously, is now shown to be a feature which is developed to a greater or lesser extent in all cells. It forms a system of membranes separating off closed channels or vesicles whose interior is not in continuity with the cytoplasm. This membranous system may be important because it provides for at least two compartments in the cell, the contents of which may be maintained under different conditions, and it may thus allow the separation of substances from each other. It also serves as a surface upon which enzyme systems may be arranged; in particular, the cytoplasmic surface of the endoplasmic reticulum appears to be a site of activity of the ribosomes, ribonucleoprotein bodies concerned in the synthesis of proteins.

The *nuclear envelope* appears to be connected with this system of

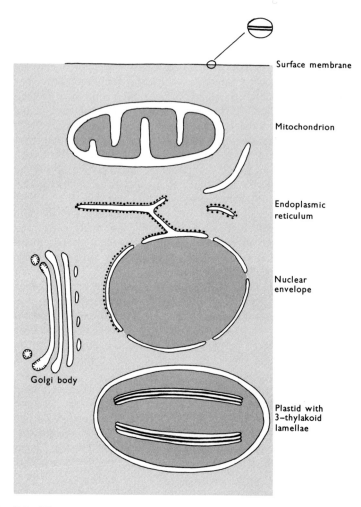

Surface membrane

Mitochondrion

Endoplasmic
reticulum

Nuclear
envelope

Golgi body

Plastid with
3-thylakoid
lamellae

Fig. 2.1 Diagrams to show the arrangement of membranes in some cell organelles; every membrane shown is a trilaminar unit membrane, as indicated at the top of the figure. Ribosome granules are shown attached to parts of the endoplasmic reticulum and nuclear membranes. The membranes of the Golgi body mature from simple vesicles which coalesce forming cisternae to 'coated vesicles' budded off from the cisternae at the extreme left.

endoplasmic reticulum membranes; indeed it often bears ribosomes on its outer, cytoplasmic surface, and may be reformed from elements of the endoplasmic reticulum following nuclear division. The two unit mem-

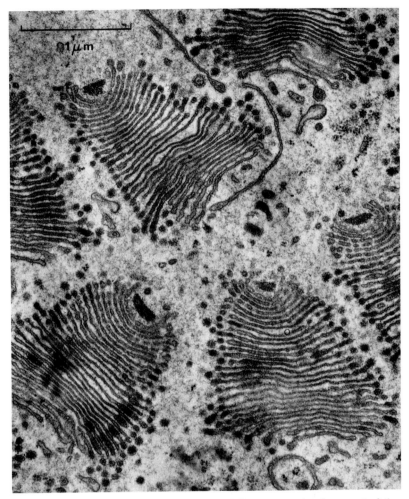

0.1 μm

Fig. 2.2 Electron micrograph of sections of dictyosomes that form part of the Golgi complex or parabasal body of the flagellate *Joenia annectens*. In each dictyosome the forming face with less distinct membranes lies against a dense body that is a section of a parabasal fibril; towards the opposite 'mature' face of the dictyosome the membranes become more distinct and characteristic dense vesicles are budded off from the edges of the cisternae. Micrograph by A. Hollande and J. Valentin.[110]

branes which comprise the nuclear envelope are separated by a space of about 20–30 nm, which is seen in some electron micrographs to be continuous with the space within endoplasmic reticulum channels. Large pores which occur in the nuclear envelope (p. 44) are supposed to provide a route for chemical interactions between the cytoplasm and the nucleus, including the regulation of gene activity and the consequent passage of gene products to the cytoplasm.

Another membranous structure associated with the endoplasmic reticulum is the cluster of flattened vesicles or cisternae which is called the **Golgi body or dictyosome** (Fig. 2.2). These are very numerous in some Protozoa, particularly flagellates, and appear to be absent or weakly developed in others (most ciliates). They are thought to be concerned with membrane maturation and with the elaboration or storage of products of cell synthesis.[191] It is frequently possible to distinguish between one (forming) face of a dictyosome at which vesicles derived from the endoplasmic reticulum or nuclear envelope aggregate to form the cisternae, and the opposite (mature) face at which vesicles are formed from dictyosome cisternae and emigrate, often to the plasma membrane. The character of the vesicles at the two faces is often different, and the cisternal membranes can often be seen to increase in thickness between the forming face and the mature face of the dictyosome (Fig. 2.2). It is thought that materials synthesized at ribosomes are transported to the forming face of the dictyosome, and within the cisternae the membrane is modified and the enclosed materials are elaborated into their final form before being transported away from the mature face. One of the best examples of this is the formation of complex scales within the Golgi vesicles of certain flagellates and amoebae, including Radiolaria, before the vesicles move to the plasma membrane and the scales are incorporated in the surface covering of the organism.[171, 172]

A more elaborate membrane arrangement is found in mitochondria and plastids, both of which are concerned with energy transformations. **Mitochondria** are usually about 1 μm across and may be 1 to several μm long. They are constructed from two membranes, the outer of which forms the surface of the structure, while the inner, which is separated from the outer by a distance of about 10 nm in most places, is continuous with projections to the interior of the mitochondrion which form lamellar, vesicular or tubular cristae. The membrane system forms two compartments; the inner membrane encloses a mitochondrial matrix compartment, and the space between the two membranes and within the projecting cristae forms a second compartment. Tubular cristae are found in most Protozoa (e.g. Fig. 2.3a). but lamellar or discoidal cristae occur in some flagellates (Figs. 5.3, 5.20), and the cristae of some amoebae and gregarines appear as vesicles rather than lamellae (Figs. 2.3b, 9.5c). The mitochondria of some Protozoa which live in environments devoid of oxygen, such as parts of the vertebrate gut, have been found to be simple sacs lacking cristae, e.g. ciliates from the rumen of cattle.[83] Among other gut-dwelling

Protozoa are some which should be regarded as facultative anaerobes and which may possess a mixture of mitochondria with cristae and simple mitochondrial sacs (e.g. *Balantidium*) or abnormal giant mitochondria (e.g. *Opalina*). Mitochondria contain the enzymes associated with the Krebs cycle and with oxidative phosphorylation (p. 60); they are the main sites of formation of the energy-rich phosphate compounds which are necessary for energy-requiring reactions in cells. Mitochondria of Protozoa and some other cells have been shown to contain DNA, and are believed to be

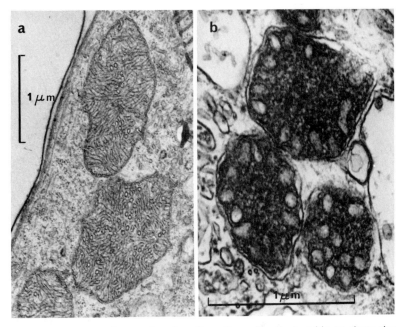

Fig. 2.3 Electron micrographs of sections through mitochondria to show the form of cristae (**a**) in the ciliate *Euplotes*, with numerous tubular cristae and (**b**) in the heliozoan *Actinosphaerium* with vesicular cristae. Micrographs by R. Gliddon and A. C. Macdonald (unpublished).

capable of the independent synthesis of their own ribosomes; at least some of the proteins required by mitochondria may depend on genes within the mitochondrion, while others depend on nuclear genes. It is thought that mitochondria may have been derived from a symbiotic association of a microorganism within another cell at an early stage in the evolution of eucellular organisms.[173]

Another class of respiratory particles are **peroxysomes**, characterized by the presence of oxidase enzymes which release hydrogen peroxide, and

catalase which oxidizes the peroxide.[18] These particles are about 0·5 to
1·0 μm in diameter and are bounded by a single membrane; in Protozoa
their presence has been established in *Acanthamoeba*[179] and *Tetrahymena*.[178]
In many cases, including *Tetrahymena*, enzymes of the glyoxylate cycle
(p. 60) are also present in the same particles, which are sometimes given

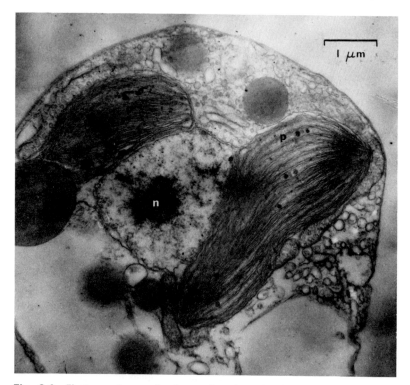

Fig. 2.4 Electron micrograph of a section through a flagellate member of the
Chrysophyceae to show the typical appearance of the chloroplast p and nucleus
n (unpublished).

the alternative name of glyoxysomes. The primary function of the glyoxylate cycle is normally the synthesis of carbohydrate from fat; in *Tetrahymena* mitochondria and peroxysomes must work together, since in this
ciliate some enzymes of the glyoxylate cycle have been found only in the
mitochondria.[178]

The **plastids** of autotrophic flagellates are also limited by two unit
membranes, but these enclose numerous lamellae within which the
photosynthetic pigments are incorporated (Fig. 2.4). In red algae the

lamellae are separate, but in flagellates there is an increased packing of the lamellae which may be grouped to form something approaching the grana which characterize the plastids of land plants. Some variants of the organization of plastids in Protozoa, and of the pigments that they contain, will be considered in Chapter 5. The plastids of some flagellates contain a pyrenoid, a region which forms a centre for the elaboration of reserve carbohydrates. Plastids have been found to contain DNA and small ribosomes capable of some independent protein synthesis; it is believed that these organelles, like mitochondria, may have been derived from symbiotic procaryotic organisms.[173]

There are also other vacuole systems with specialized functions, notably the contractile and food vacuoles, which deserve some comprehensive treatment here because they are best known in Protozoa.

Contractile Vacuoles[140, 143]

Contractile vacuoles are present in most ciliates and in many flagellates and amoebae; they are probably most conveniently studied in ciliates, particularly sedentary forms. Many Protozoa have several contractile vacuoles, others have only one. These vacuoles occur close to the plasma membrane, and are seen to swell slowly (diastole) before suddenly collapsing (systole) and releasing their fluid contents to the outside medium. In many Protozoa, particularly those with a specialized body surface, the contractile vacuoles have fixed positions; they are filled from systems of canals, and open to the outside through permanent pores. In others, mainly amoebae, the contractile vacuole has a less organized structure—it may increase in size by the coalescence of smaller vacuoles, and may be carried about in the cytoplasm before release at the body surface.

In electron micrographs of the simple contractile vacuole of *Amoeba*,[177] the unit membrane enclosing the clear vacuole is surrounded by a region containing many small vesicles and mitochondria (Fig. 2.5a). Some of the small vesicles may be seen to have direct continuity with the vacuole, so it appears that small vesicles form in the cytoplasm and discharge into the contractile vacuole during diastole. The expulsion of fluid at systole of this type of vacuole is rather feeble, and no good evidence is available of any fibrous system that could be contractile.

A much more complex structure has been found in the contractile vacuole of *Paramecium*[129, 226] (Fig. 2.5b), where the main vacuole is fed by about six radiating canals. Diastole of the main vacuole is achieved by systole of the canals, and only during this time is there an open connection between the ampullae at the proximal end of the canals and the main vacuole; fluid does not flow back into the canals during systole of the main vacuole. The canals are surrounded by a 1 μm thick layer of fine branching tubules about 20 nm in diameter, which communicate at their outer ends with tubular components of the endoplasmic reticulum; at their inner ends the tubules open to the vacuolar canals during diastole, but during systole of the canals the connections between tubules and canals are closed (Fig. 2.5c, d). There is evidence to suggest that pressure exerted on

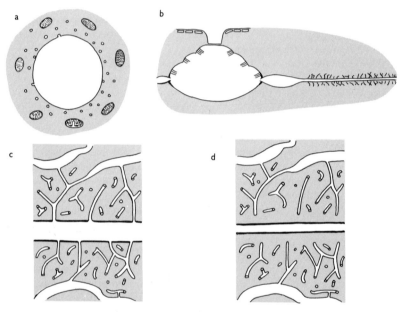

Fig. 2.5 Diagrams to show the structure of contractile vacuoles. (**a**) A section of the contractile vacuole of *Amoeba* surrounded by vesicles and mitochondria (drawn from information published by E. Mercer).[177] (**b**) A section through the contractile vacuole and parts of two supply canals of *Paramecium*. Fine tubules surround the canal, which widens at the ampulla before opening into the main vacuole; this is surrounded above by bands of fibrils, and periodically opens to the surface at a special pore through the pellicle. Connections between the fine tubules and the canals of *Paramecium* are open during diastole of the canals, (**c**), but are closed during systole of the canals (**d**) (drawn from information published by L. Schnieder).[226]

the vacuole by the cytoplasm is responsible for the expulsion of fluid at systole; bundles of fibres which surround the ampullae and the main vacuole are assumed to be concerned in the production or control of the pressure required to contract the vacuole. Clearly the contractile vacuole apparatus involves not only the vacuole visible in the light microscope, but also a considerable area of the surrounding cytoplasm. Comparable structures involving a spongy region with feeding tubules and a system of fibres around the vacuole have been reported in other ciliates.

The activity of contractile vacuoles is normally much higher in freshwater forms than in marine Protozoa and endosymbionts; this observation, together with experimental evidence, suggests that the primary function of the contractile vacuole is osmotic regulation. There is evidence that the osmotic pressure of the cytoplasm of freshwater Protozoa is well above that of the surrounding medium, and it is likely that marine Protozoa are slightly hypertonic to seawater. As a result water will flow by osmosis into the bodies of Protozoa. Water will also enter with food taken in by phagocytosis or pinocytosis, and may be produced in metabolism. Variation in the output of a contractile vacuole, as required by changes in water content of the body, may be achieved by change in volume of the vacuole or by

Table 2.1 Comparative data on the contractile vacuole activity of several Protozoa.

Species	Duration of one cycle seconds	Rate of output μm^3/sec	Time to expel body volume minutes	Temperature °C	Medium
Amoeba proteus[140]	150–800	54–109	230–800	19–27	Hay infusion
Paramecium caudatum[140]	6–30	54–258	15–49	15–23	Culture medium
Carchesium aselli[140]	6–39	6–20	25	14·5–16	London Tap Water
Discophrya collini	25–100	3–15	25–120	20–25	Bristol Tap Water
Cothurnia curvula[140]	30–1800	0·1–1·7	240	14·5–16	Sea Water
Acanthamoeba castellanii[192a]	mean ~50	mean 2·1	15–30	24–25	1% Peptone

change in the duration of the vacuolar cycle. The rate of pulsation of the vacuole varies in different Protozoa (Table 2.1); in some of the smaller freshwater ciliates the vacuole may contract every few seconds, and in some larger endo-symbiotic ciliates the vacuole may contract only once or twice in an hour. The relative rate of removal of water also appears to be greatest in the smallest forms, e.g. a small ciliate may expel a water volume equal to its body volume in a few minutes, while a large ciliate may take several hours to pump out the equivalent of its body volume. Some typical data are given in Table 2.1.

Changes in the rate of vacuolar output of the particular organism may be caused by alterations to the concentration of osmotically active solutes in the surrounding medium. J. A. Kitching[139] found that when the suctorian *Discophrya* is placed in a dilute solution of sucrose (to which the body surface is not permeable) the rate of output of the vacuole adjusts to a new level after a short lag period, and returns to the original rate if the suctorian is returned to water (Fig. 2.6a); under these conditions the vacuolar output of the organism is proportional to sucrose concentration and the body size remains the same (Fig. 2.6b). At a sucrose concentration of about 50–60 mM

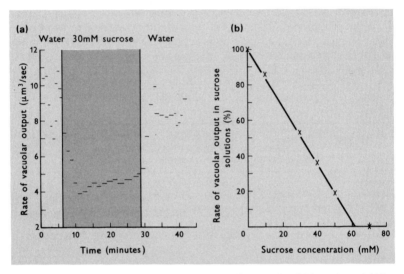

Fig. 2.6   Results of experiments on the contractile vacuole of *Discophrya*. (a) The rate of vacuolar output, calculated from the diameter of the full vacuole and the duration of the vacuolar cycle, in water, during a period when the suctorian was surrounded by 30 mM sucrose solution and following the return to water; each short line represents the rate calculated for one vacuolar cycle. (b) A comparison of the concentration of sucrose solutions with the rate of vacuolar activity in those solutions expressed as a percentage of the rate in water (unpublished).

the vacuole ceases to contract, and if the sucrose concentration is increased further the body shrinks. It is assumed that the body contents are isosmotic with the sucrose solution which just causes cessation of the vacuolar pulsation. In solutes which penetrate the body surface (e.g. ethylene glycol), a reduction of vacuolar rate on immersion in the solute is followed by a gradual return towards the resting level as the solute concentration inside the body rises: similarly, on return to the original medium the activity of the vacuole is increased temporarily until excess solute has been removed from the cytoplasm. Similar changes occur if *Discophyra* is transferred to dilute seawater and then returned to freshwater.

The means by which the contractile vacuole exerts an osmoregulatory control are not entirely clear. It has been found by freezing point determination on fluid from the contractile vacuole of *Amoeba* that the osmotic concentration of the vacuolar fluid was about $\frac{1}{3}$ of that of cytoplasm and about 4 times that of the outside water.[225] Because the vacuolar fluid is hypotonic to the cytoplasm, and because of the rapid throughput of water in many organisms, it is believed that some active transport of water into the vacuolar system or of solutes out of the vacuolar fluid must occur. It is often assumed that metabolic products and waste nitrogenous substances are removed from the body in the vacuolar fluid; however, while urea has been found in the contractile vacuole fluid of the large ciliate *Spirostomum*, it is likely that most nitrogenous wastes of Protozoa are liberated as ammonia and diffuse rapidly away.

The dinoflagellates have vacuoles called pusules, sometimes formed into complex systems (p. 129), whose function has been thought to be the same as that of contractile vacuoles.[61] However the pusules only show occasional swellings and shrinkage and do not undergo regular contractions. In complex pusule systems there is a main vacuole surrounded by subsidiary vacuoles in a specialized region of cytoplasm and connected to the surface by one or more tubular canals.[62]

Food vacuoles[141, 248]

Another important type of vacuole is that which is concerned with the intake and digestion of food. Food vacuoles are conventionally divided into phagocytic vacuoles which enclose large food particles and pinocytotic vesicles which enclose invisible food materials either in solution or adsorbed on the surface membrane. It does not seem possible to make a sharp distinction between these two types of vacuole.[198] In many Protozoa with well-formed cytostomes the intake of food may only involve phagocytic vacuoles, and in many other forms with no permanent cytostome pinocytosis may be the main mode of feeding; in the majority of Protozoa, however, it is likely that both types of feeding occur. It is likely in any case that pinocytosis may occur at the surface of phagocytic vacuoles. The development of invaginations, which may be cut off by the cytoplasm to form pinocytotic vesicles, has frequently been observed in amoebae; the rate of formation of the vesicles may be stimulated by protein in the medium, but not by carbohydrates. Phagocytic vacuoles are formed following food cup formation by amoebae and other forms which have a very flexible surface, the cup being formed by pseudopodial extensions of the body which flow over and around the prey organism or food particle and eventually enclose it. Similar food vacuoles are formed at the cytosome of ciliates where the presence of food induces the formation of an invagination to enclose the food collected by the activity of cilia or other feeding structures. Phagocytic vacuoles may also surround certain organelles prior to their autodigestion during morphogenetic reorganizations of the body.

The organic contents of these food (phagocytic and pinocytotic) vacuoles are digested by the activity of enzymes contributed by lysosomes, which are presumably the neutral red granules of earlier protozoologists (Fig. 2.7). Lysosomes are bodies about 0·2–0·8 μm in diameter which are believed to be formed in the Golgi region; they have a single unit membrane and contain a collection of hydrolytic enzymes, with a common (acid)

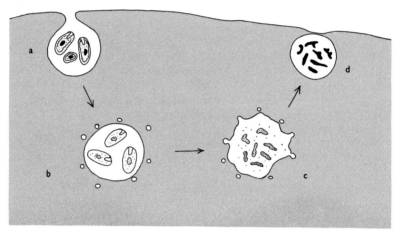

Fig. 2.7 Stages in the life of a food vacuole. Prey organisms taken into a vacuole by phagocytosis (**a**) are killed and digested by the contents of lysosomes that coalesce with the food vacuole (**b**); following digestion micropinocytosis may be seen around the vacuole (**c**) and finally the undigested remains are taken to the body surface (**d**) for removal from the body.

optimal pH, capable of the breakdown of the major classes of organic molecules to simple soluble molecules of the order of size of amino acids and monosaccharides.[284] Generally food vacuoles shrink slightly soon after formation, and then become acid at about the time that the prey organism is killed—in *Paramecium* pH values between 1·4 and 3·4 have been reported. Later the contents of the vacuole become alkaline, and the vacuole swells again, presumably because of the appearance in it of the products of digestion of the prey. Subsequent absorption of soluble materials into the cytoplasm, probably accompanied by micropinocytosis around the surface of the food vacuole, leads to a final shrinkage of the vacuole to a bag containing undigestible residue which leaves the body through the unspecialized surface membrane or at a cytopyge in organisms which have a specialized pellicle.

FIBROUS STRUCTURES

The cytoplasm of Protozoa also contains fibrous structures, the majority of which can be assigned to one or other of two classes—those built of filaments 4–10 nm thick or those built of microtubular fibrils about 20 nm in outside diameter.[238] Commonly the *filaments* appear to be built of

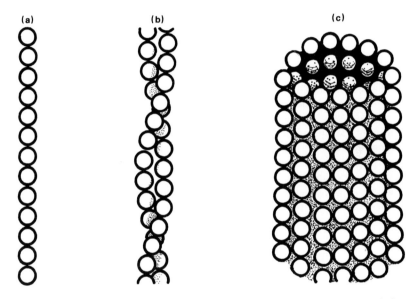

Fig. 2.8 The arrangement of globular molecules in filaments and microtubular fibrils. The single row of molecules (**a**) seems to be less frequent than the filament composed of two strands (**b**), an arrangement found in muscle actin; the wall of the cylindrical microtubule (**c**) contains many rows of molecules.

single rows of globular protein molecules (Fig. 2.8a) or of two or more of these molecular chains woven together (Fig. 2.8b). In fibres that are capable of change in length it is believed that the component filaments are loosely aggregated into bundles, while in more rigid fibres responsible for resisting changes in length the filaments are accurately lined up and cross-linked with regular periodicity to form the striated fibres that are often seen in electron micrographs. Examples of the loose bundles are found in the contractile myonemes of the ciliates *Spirostomum* and *Stentor* (Fig. 8.4 p. 191), and in the contractile spasmoneme of the stalk of *Vorticella*.

Striated fibres constitute the kinetodesmata of such ciliates as *Paramecium*, and some of the roots originating from the basal bodies of flagellates are striated (Fig. 5.30 p. 145).

Microtubular fibrils appear to be constructed from a number of uniseriate rows of globular protein molecules joined together to form the walls of the cylindrical fibril (Fig. 2.8c). In several studies of fibrils with a diameter of about 20 nm it has been reported that 12–13 rows of molecules are present; it is possible that larger or smaller microtubular fibrils have a larger or smaller number of component rows. Such microtubular fibrils form the basis of a number of fibrous structures in Protozoa of all types. They occur in the fibrous axis of the arm of heliozoan amoebae (p. 160), in the rods associated with the mouth of flagellates like *Peranema* (p. 67) and ciliates like *Nassula* (p. 67); they abound in the pellicles of flagellates (e.g. *Trypanosoma*, p. 140), sporozoan trophozoites (e.g. *Selenidum*, p. 229), and ciliates (e.g. *Euplotes*, p. 193); they form the longitudinal elements of flagella and cilia and occur in other situations, many of which are mentioned elsewhere in this book. While microtubules appear in a very wide variety of places, and may be involved in a variety of functions, there is an increasingly widespread belief that they are rigid fibrils, incapable of changes in length (except by addition or removal of material), so that if they are involved in movement this must be achieved by the fibrils sliding along one another; a possible exception that deserves detailed investigation is the axopodia of some centrohelids (p. 163) and Radiolaria (p. 162). Microtubules also resist transverse distortion, and are often cross-linked in bundles to form very rigid structures.

Structure of cilia and flagella[232, 237, 239]

Cilia and flagella are characteristic organelles of a great many Protozoa and deserve detailed description. The two types of organelles are structurally the same, but have minor functional differences; it is probably reasonable to describe cilia as a specialized class of flagella. Flagella are cylindrical organelles about 0·25 μm in diameter, composed of a longitudinal bundle of microtubular fibres (the axoneme) enclosed within a unit membrane which is continuous with the plasma membrane of the cell. The axoneme extends into the surface region of the cell as the flagellar basal body, sometimes called the blepharoplast in flagellates or the kinetosome in ciliates.

The microtubular fibrils of the **axoneme** are arranged in a precise pattern, best seen in cross sections, in which two microtubules 24 nm in diameter, with a wall thickness of about 4 nm and spaced with 30 nm between centres, occupy the centre of the bundle and are surrounded by 9 double microtubular fibrils arranged on the circumference of a circle about o·2 μm in overall diameter (Fig. 2.9). The peripheral doublet fibrils measure about 37 nm in the tangential direction and 25 nm radially, the common wall between the two microtubules being the same thickness

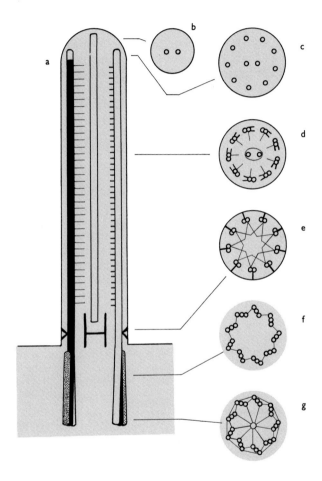

Fig. 2.9 Diagrams showing the structure of a flagellum. A longitudinal section is shown at the left (**a**), and (**b**) to (**g**) show transverse sections taken at the levels indicated along the length of the flagellum (details are described in the text).

(about 4 nm) as the outer walls. All of the fibrils appear to run straight along the flagellum without spiralling or twisting. The complete microtubule (subfibril A) of each doublet carries two rows of lateral projections (arms), about 8 nm thick and 14 nm or more long, which are directed towards the adjacent doublet, and which occur at intervals of about 17 nm within each row along the long axis of the doublet fibril.

The central microtubular fibrils appear to be constructed from 13 rows of molecular subunits, and about 23 rows of subunits compose the doublet fibrils.[212] The subunits of both types of fibril are globular protein (tubulin) molecules about 4 nm in diameter and with a molecular weight of about 55 000. The solubility properties of the central, peripheral A and peripheral B fibrils differ, and R. E. Stephens has found that there are minor differences in amino acid composition of the tubulin protein of the A and B fibril components in sea urchin sperm flagella.[246] Certain non-motile ('paralysed') mutant strains of such flagellates as *Chlamydomonas* have been found to lack the central fibrils, but possess a full set of peripheral fibrils; this indicates both that flagellar protein production is under genetic control and that the production of proteins for central and for peripheral fibrils is under separate control.[282] There is evidence that the tubulin proteins are of the actin type, well known in muscle, but that in flagella the molecules polymerize in a different configuration and carry bound guanosine nucleotides instead of the adenosine nucleotides associated with muscle actin. Research on this protein has utilized mainly cilia from *Tetrahymena* and flagella from sea urchin spermatozoa; the protein now known as tubulin has also been called flactin or spactin by some authors.

The arms on peripheral doublets of the axoneme are composed of much larger molecules of a protein called dynein,[85] which is of a similar size to muscle myosin (MW ~ 500 000), but has an ATPase which is activated by both magnesium and calcium ions, and has a very different shape from the myosin molecule of striated muscle. Dynein molecules isolated from cilia are the shape of the arms (14 nm long and 8 nm thick), and are often found polymerized in chains in which adjacent arm units appear to be linked by a slender thread; two of these chains are presumably bound to the A subfibril of each doublet.

Other structural components of flagella are regularly seen. Subfibril A of each doublet carries an additional projection which extends radially towards the centre of the cilium and appears in some preparations to be knobbed at the end. It is not certain whether this radial strand might make contact with some central structure in the intact cilium or whether the knobs are the remnants of secondary longitudinal fibrils that have been described in a position midway between the central and peripheral fibrils of some flagella. The two central fibrils commonly appear to be surrounded by a membrane or perhaps a spiral thread, and radial strands could connect with this. It is not yet clear whether all the components of the axoneme described by F. Warner for insect sperm flagella occur in all flagellar organelles.[281]

At the flagellar tip the central fibrils may be longer than the peripheral fibrils, so that the flagellum has a narrowed terminal region; in other cases the fibrils may be more nearly the same length. Commonly the peripheral doublets become single near the tip by the termination of the B subfibril; the arms and radial strands do not occur where the peripheral fibrils are single.[223]

The structure of the *basal body* has some consistent features, but varies widely in details of fibrous connections within and outside the axonemal fibril complex (Fig. 2.9). At the transition between the flagellum and the basal body the arms and radial strands on the peripheral doublets are missing, and frequently links are seen between the doublets and the flagellar membrane. The central fibrils end at about the level of the cell surface, and may be embedded at their inner ends in a granule or transverse plate which may fill much of the area within the peripheral ring. Between the termination of the central fibrils and the transverse plate of the flagella of several flagellates a specialized zone has been discovered, in which a filament forms a stellate pattern within the ring of peripheral doublets, making contact with alternate members of the ring as it spirals along the flagellar axis.[169] Within the basal body a third microtubular subfibril (C) is added to each doublet, and the triplet is twisted so that the A subfibril lies nearer the centre and the C subfibril further from the centre than the B subfibril. Thin strands inter-connect the triplets in various ways, the commonest being the cartwheel arrangement in which each A subfibril is connected to the adjacent C subfibril, all C subfibrils are connected around the circumference and from each A subfibril a strand connects with a central ring. The central area is sometimes occupied by granules or other structures.

The flagellar basal body has the same detailed structure as the *centriole* found in animal cells. In some flagellates the basal bodies act as the organizing centres for the mitotic spindle of the dividing nucleus, functioning as centrioles and bases of motile flagella at the same time.[21, 167] The centriole may have been derived from the flagellar basal body of ancestral flagellates. The basal body is also the structure from which flagella and cilia develop, and an organizing centre for many cytoplasmic fibril systems. There is some evidence that basal bodies contain DNA and may have some powers of self replication. It has long been believed that basal bodies only developed from or in close association with existing basal bodies.[162] Complete proof of this is not available, and in some cases it has been doubted whether basal bodies persist through some life cycle stages from which flagella are absent, since no centrioles have been found in the amoeboid stages of *Naegleria* in spite of careful fine-structure studies.[57, 80a] In several organisms a short basal body or centriole has been seen at an early stage of development lying close to the proximal (inner) end of an existing basal body or centriole, and with its long axis at right angles to the long axis of the complete basal body.[3] This is assumed to be the normal site of formation of basal bodies; there is no evidence that a basal body ever divides, but rather that a new organelle is normally formed very close to an existing one.

Fibrillar 'rootlet' structures associated with basal bodies and extending away into the cytoplasm include groups of microtubular fibrils, striated fibres made up of filaments, and sometimes non-striated filamentous fibres;[200] these vary widely in different forms, and examples will be mentioned in later descriptions of some Protozoa. Generally the proximal ends of these fibres are embedded in electron-dense material which also extends around some of the triplets of the basal body.

Cilia are frequently combined together to form compound organelles called cirri, membranelles or ciliary undulating membranes; no membranes or formed structures seem to be present to hold the component organelles together, although adhesion by secreted mucous material is a possibility. A few structures in Protozoa are derived from modified ciliary organelles, e.g. the outer region of the stalk of *Vorticella*, which develops from a scopula, but in motile cilia the structure is extremely constant except for the different lengths of organelles from different places. By contrast, there are several interesting variants of the flagellar pattern. They include intra-flagellar material lying alongside the axoneme; hairs and scales or other structures attached to the flagellar membrane; and extensions of the flagellar membrane which may form a flagellar undulating membrane, one edge of which is attached to the surface of the organism while the other edge contains the axoneme. These flagellar variants are generally charac-teristic of particular groups of flagellates, and will be described in more detail later (Chapter 5). While additional intraflagellar material may be important in providing additional stiffness within the flagellum, the external structures have an effect on the hydrodynamics of flagellar locomotion.

Haptonemata

Certain flagellates carry an organelle called a *haptonema*,[195] which is similar in dimensions to a flagellum but which has a different internal structure and moves in a different manner. At its extreme base the haptonema is first seen in cross sections as a group of about 9 micro-tubules,[168, 170] in approximate hexagonal array, lying between the two flagellar bases (Fig. 2.10). At higher levels towards the body surface the number of fibrils is first reduced to 8 (arranged in two concentric arcs of 4 fibrils) and then to 7, which arrange themselves into a C-shaped row at the level of the cell surface, and are eventually formed into a loose ring at high levels within the free organelle. As few as 5 or 6 microtubular fibrils have been found in some haptonemata. Surrounding the fibrils of the organelle are three concentric membranes, the outermost of which is continuous with the plasma membrane, while the two inner membranes enclose a vesicular cavity which extends around the haptonema throughout its length and also penetrates a short distance into the body of the flagellate. During development this vesicle is continuous with the endoplasmic

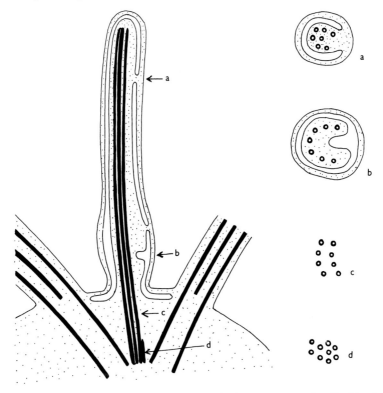

Fig. 2.10 Diagrams showing the structure of a haptonema, as described in the text. Based on electron micrographs by I. Manton and her colleagues.[168,170,195]

reticulum of the cell. A basal swelling of the haptonema has been found in *Chrysochromulina*, and in this form and *Prymnesium* a vesicular process extends across a large part of the cross sectional area at the base of the shaft of the organelle; the significance of these structures in the functioning of the organelle is not known. Haptonemata move by coiling up and uncoiling; the fully coiled organelle appears to be in the relaxed position, and B. Leadbeater has described movements in which the extension by uncoiling is slow while the retraction with coiling is more rapid.[147] The name haptonema (adhesive thread) was chosen to describe a feature of these organelles, which is that they cause adhesion of the organism to foreign objects, but the significance of this property in the life of the flagellate is not known.

Movement of cilia and flagella[112, 235]

Protozoa have been used frequently in studies of the movement of flagella and cilia and of the mode of co-ordination of cilia; they provide a rich variety of examples. Since these organelles beat very quickly (often more than 20 times per second), it is necessary to employ special methods for studying them. The simplest apparatus is a stroboscope, which may be a mirror or perforated disc rotated by an electric motor, or an electronic flash tube driven by a timing circuit; in either case the flashes of light are made to occur at the same instant in each cycle of beat of the organelle, which therefore appears to stand still. This technique allows the study of the metachronal waves of cilia, and if the stroboscope is set at a slightly lower frequency than a cilium the form of beat of the slowly moving organelle may be studied. High speed cine-photography is also a most valuable tool in the study of the movement and of the co-ordination of cilia, and prints from such films allow detailed analysis of the form of beat of organelles and of the inter-relations between neighbouring cilia participating in a metachronal wave. Photographs taken with electronic flash may be used to supplement either of these techniques.

The movement of external motile organelles may be slowed down considerably by the use of such viscous agents as methyl cellulose; in these media flagella and cilia show an abnormal beat, and cilia show modified co-ordination patterns, so that observations must be interpreted with extreme caution. The shape of cilia at various stages of their cycle of beat and the form of metachronal waves have been studied on 'instantaneously-fixed' ciliated Protozoa (and *Opalina*) with the light microscope and the scanning electron microscope, using the technique of fixation described by B. Parducz.[193, 253] In this case also it is necessary to insist on cautious interpretation of the pictures obtained, although H. Machemer has shown that profiles of cilia seen in flash photographs of living *Paramecium* appear the same as those of cilia on instantaneously fixed specimens.[164] For stationary organisms with a regular ciliary beat the stroboscope remains the most convenient aid in observation, but for actively motile organisms or those with irregularly beating organelles the use of cine-photography is most useful, and flash photography is a valuable aid.

Flagella move in an undulating manner (Fig. 2.11). The movement may take place approximately in a single plane, so that the envelope of movement is almost a flat rectangle, or the flagellum may move out of the plane to a greater or lesser extent, so that the envelope of movement becomes more nearly cylindrical as the beat tends towards the helical form. Such movements are normally regular and symmetrical in a healthy flagellum, and several complete waves of movement of the organelle may frequently be seen within the length of the flagellar shaft. In some cases the movements towards either side become asymmetrical, and an extreme form of such a unilateral tendency is the ciliary beat described below. Helical or planar undulations may travel along the flagellar shaft in either direction (Fig. 2.12) but the waves of movement originate at the flagellar base much more commonly than at the tip of the flagellum.[111, 112] The movement of the flagellum produces a resultant force on the water acting along the long axis

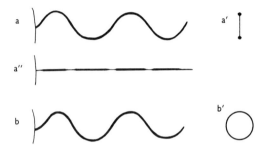

Fig. 2.11 The movement of flagella. In some flagella the propagated waves are planar and the flagellum appears different when viewed in different directions (a in side view and a″ with the flagellum viewed in the plane of the waves,) while in other flagella the waves are helical and appear the same when viewed from different angles around the flagellum, b. The appearance of the two types of waves as seen from the flagellar tip is shown at a′ and b′.

of the flagellum in the direction of the flagellar waves, provided the flagellum has a smooth surface. In considering the effective propulsive force of a flagellum it is important to realize that inertial forces are negligible in comparison with viscous forces, in structures as thin as a flagellum undulating many times a second in water. Part of the available energy must be used up in changing the shape of the flagellum (i.e. in overcoming internal elastic resistance) and part is used in doing work against the viscous resistance of the water around the flagellum. It has been calculated that under optimal conditions the viscous work done by a flagellum is about three times the elastic work, and C. Brokaw calculated that the total energy dissipation of a 40 μm long flagellum beating about 30 times a second with a wavelength of about 30 μm is about 3×10^{-7} erg/sec at 16°C.[25] The locomotive force produced by a given flagellar waveform depends principally on the amplitude and wavelength of the waves and the speed at which these waves move along the flagellum; the dimensions of the structure are of rather less importance. The viscosity of the medium is only important in that it may affect the shape of the waves and their frequency.

Lateral hairs on the flagellum modify the movement of water that is produced by the flagellar undulations. Lateral hairs are of at least two types; long slender hairs about 5 nm thick, frequently in one row, as in

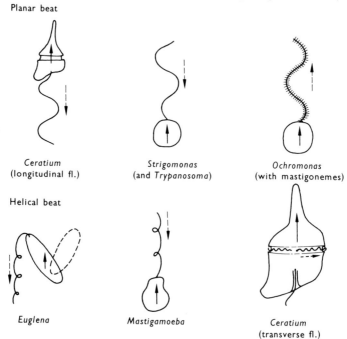

Planar beat

Ceratium
(longitudinal fl.)

Strigomonas
(and *Trypanosoma*)

Ochromonas
(with mastigonemes)

Helical beat

Euglena

Mastigamoeba

Ceratium
(transverse fl.)

Fig. 2.12 Diagrams to illustrate some of the ways in which flagella may be used to cause locomotion of Protozoa. In each case the arrows with broken lines indicate the direction of propagation of flagellar waves and the full arrows indicate the direction of locomotion.[238]

Euglena, and thicker, stiffer projections about 20 nm thick and 1 μm or so long (Fig. 2.13). It is usual to refer to all types of flagellar hairs as flimmer filaments and to reserve the term mastigoneme for the thicker projections. Mastigonemes are believed to project rigidly from the flagellum in two rows arranged in the plane of the flagellar undulation; they cause a reversal of the flow of water produced by the undulation of the flagellum so that the propagation of bending waves along the flagellar axis from base to tip results in a flow of water towards the flagellar base (Fig. 2.12 and

Fig. 5.11c, p. 122). The observed rate of water movement produced by a chrysomonad flagellate bearing mastigonemes was of the same order as the rate calculated by M. E. J. Holwill, using hydrodynamic equations

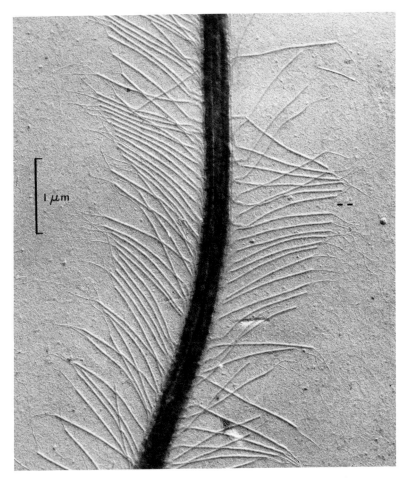

Fig. 2.13 Electron micrograph of a part of a flagellum of *Paraphysomonas* showing the form of the thick lateral hairs, which have three regions, the basal part, the main shaft and the fine bristles at the tip (see also Fig. 5.6). Micrograph by B. S. C. Leadbeater.[148a]

involving surface coefficients of resistance to movement of the flagellum and its mastigonemes.[113]

The use of the techniques of observation mentioned above has also shown a rich diversity of patterns of ***ciliary beating***.[235] Frequently the ciliary beat takes place more or less in a single plane, and involves two phases, one in which the cilium moves towards one side while bending only in the basal region, and a second in which the region of bending is propagated up the ciliary shaft to the tip and returns the cilium to the starting position (Fig. 2.14a). The latter phase usually occupies much more than half of the beat cycle, and is referred to as the recovery stroke, while the rigid swing is referred to as the effective stroke in recognition of the fact that the movement of water achieved by this stroke is normally much greater than that in the recovery stroke. The 'effective' stroke has been found to occupy more than half the cycle in some cilia, and in at least some cases it moves

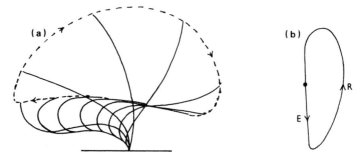

Fig. 2.14 The movement of cilia. A series of profiles showing an unspecialized example of a cycle of ciliary beating as viewed from the side (**a**) may be subdivided into the effective stroke during which the ciliary shaft moves from left to right and swings through a large arc, and the recovery stroke during which the bent cilium unrolls towards the left. The path followed by the ciliary tip sometimes follows an anticlockwise loop, as shown at **b**, where the cilium is viewed from the tip and part E represents the effective stroke and part R the recovery stroke.

less water than the 'recovery' stroke; these cilia are believed to be specialized for some function other than the propulsion of water. Where the beat is not planar it appears that the more rapid movement of the effective stroke is approximately in one plane, while the movement in the recovery stroke involves a swing of the cilium to one side. In Protozoa this swing appears to be consistently to the left of the direction of the active stroke, so that viewed from the ciliary tip the recovery stroke moves anticlockwise[193, 164] (Fig. 2.14b), but a swing to the right occurs in some Metazoa.[1]

The analysis of high speed cine films has provided data for calculations of the work performed against viscous resistance during the ciliary beat,[240] and of the stiffness of the ciliary shaft at different parts of the cycle of beat.[211] The viscous work done during the effective stroke, and thus the propulsive effect on the water, is dependent on the cube of the ciliary length and the square of the angular velocity of the beat; the viscosity of the medium and the duration of the stroke are also important, but the ciliary diameter is of less importance. In one example,[240] the work done against viscosity in one effective stroke lasting 24 msec by a cilium 32 μm long was calculated to be about 4×10^{-9} ergs. In the recovery stroke of the same cilium (lasting 36 msec) the viscous work done was calculated to be about 1×10^{-9} ergs. The work done against the stiffness of a cilium depends on the amount of bending (the radius of curvature), on the modulus of elasticity of the material of the structure, its cross sectional area and distance from the axis of bending. There is some evidence that the stiffness of the ciliary shaft is some 10 times less during the recovery stroke than during the effective stroke, the additional stiffness of the latter being provided by active forces within the cilium.[211] On this basis it is estimated that the cilium mentioned above, which performed about 5×10^{-9} ergs of viscous work per cycle, would perform between 10^{-9} and 10^{-8} ergs of elastic work against the stiffness of the cilium in each cycle. This requires a total power dissipation by the cilium of between 10^{-7} and 10^{-6} ergs/sec, which is closely comparable with the power output of a flagellum.

The co-ordination and control of cilia and flagella[137,230,234,237]

When a cilium moves it carries with it a surrounding layer of fluid, the extent of which depends on the viscosity of the fluid and the velocity of the cilium. In the normal pattern of ciliary beat, therefore, more water is carried by the faster effective stroke than by the slow recovery stroke, so that a pulsed flow of water is caused from one side of the cilium towards the other. When two cilia lie close enough for their transported water layers to overlap, interference will occur between the movements of the two cilia, and the activity of the two cilia will probably become hydrodynamically linked. Such hydrodynamic linkage will also occur between two moving flagella which lie close together; this accounts for the synchrony of beat seen in tufts of flagella, and the co-ordination between the flagella of separate spermatozoa seen in packed suspensions. It is also assumed to be responsible for the waves of co-ordination of the flagella which densely cover the bodies of some of the larger zooflagellates, and even for the co-ordinated movement of spirochaetes that live attached to the large flagellate *Mixotricha*.[35]

The form of interaction between two adjacent cilia will depend on the positional relationship between them (distance and position with respect to the plane of beat), on their length, and on the pattern of ciliary beat that they perform. In most cases the co-ordinated activity of cilia is held to result from hydrodynamic linkage of this type. The compound ciliary organelles of Protozoa contain cilia which are structurally separate; they appear to beat as a single compound unit in life, but fray apart quickly on death. The bases of the component cilia in cirri and membranelles lie so close together that there is tight viscous-mechanical coupling (=hydrodynamic linkage) between adjacent cilia, and the beat of all of the units is normally synchronized unless the compound structure is artificially split apart. In a ciliary undulating membrane (p. 199) the basal bodies of the cilia lie very close together, but are arranged in a long line; if a cilium at one end of the line beats, the motion of this cilium is coupled to the next

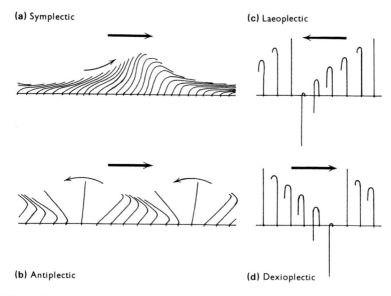

(a) Symplectic

(c) Laeoplectic

(b) Antiplectic

(d) Dexioplectic

Fig. 2.15 Diagrams to show the relation between the direction of the effective stroke of the ciliary beat and the direction of movement of metachronal waves in the four main patterns recognized and named by E. W. Knight-Jones.[144] In c. and d. the effective stroke of the ciliary beat is towards the observer.

and is passed to all members of the row in sequence, so that the cilia beat metachronally and waves of movement pass along the membrane. In many rows of cilia the bases are further apart, and the cilia do not form a membrane, but they may still beat in metachronal waves if there is sufficient hydrodynamic coupling between them.

The patterns of ***metachronal co-ordination*** which result are diverse (Fig. 2.15). In long rows of cilia the plane of the effective stroke of the beat is usually approximately at right angles to the row along which the metachronal waves pass, and the metachronism is referred to as diaplectic, conforming to either the dexioplectic pattern when the beat is towards the right or the laeoplectic pattern when the beat is towards the left of the line of propagation of the waves. A few examples are known in which the metachronal waves of cilia pass over a field of cilia in the same direction as the effective stroke of the ciliary beat (symplectic metachronism) or in a direction more or less opposite to the effective stroke (antiplectic metachronism). The dominant coupling between cilia in many examples of both antiplectic and diaplectic metachronism seems to take place in the recovery stroke of the beat, and the increased spacing between the cilia during their effective stroke permits a faster movement during this stroke. The metachronism of *Paramecium* is approximately dexioplectic (according to the description of H. Machemer[164]), but it is possible to modify the coupling between cilia during the effective stroke by adding a viscous agent such as

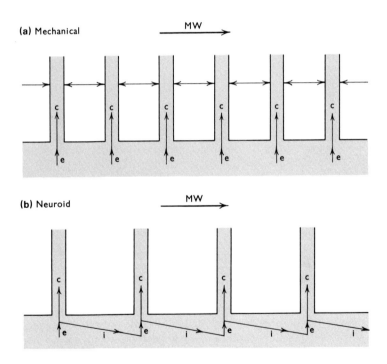

(a) Mechanical

(b) Neuroid

Fig. 2.16 Diagrams illustrating two possible mechanisms of metachronal co-ordination. The excitation (e) that leads to contraction (c) may be timed as a result of viscous-mechanical constraints acting continuously between the ciliary shafts in mechanical co-ordination (a), or may be timed by some form of internally mediated excitation-coupling (i) in neuroid co-ordination (b).

methyl cellulose to the medium around the ciliate, and thereby to induce near-symplectic metachronism.

The co-ordination between cilia that results in the formation of metachronal waves is normally achieved by viscous-mechanical coupling between adjacent active cilia, whose mutual interaction maintains a phase difference between the cilia in the direction of wave propagation, according to the asymmetry of the ciliary beat (Fig. 2.16a).

The metachronism of the membranelles (compound cilia) of *Stentor*

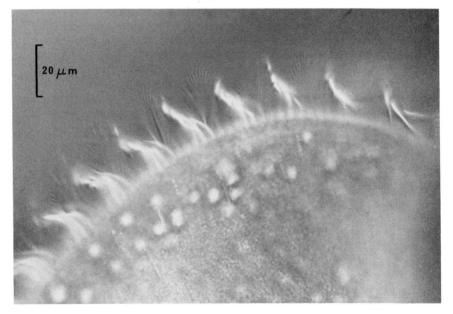

Fig. 2.17 Metachronal waves of membranelles of the ciliate *Stentor* (see p. 212). The fanned out compound cilia move towards the observer in the effective stroke and the waves move to the right.

(Fig. 2.17) does not appear to be of this type, because the movement of the waves appears to be largely independent of the rate of beat of the cilia.[234] It is suggested that the coupling between adjacent membranelles which maintains the metachronal co-ordination is a function of the basal cytoplasm, and that only in media of relatively high viscosity, more than three times normal, is there sufficient hydrodynamic coupling between the membranelles to be responsible for metachronism. There is evidence that the internal co-ordination process does not involve a continuous conducting fibre underlying the membranelle row, but rather that the beat of one membranelle produces some change in the cytoplasm which contributes to the excitation of the next membranelle; the propagation of metachronal waves requires the active participation of the membranelles (Fig. 2.16b).

Many ciliates and *Opalina* are able to change the direction of movement by alteration of the *direction of beat* of the cilia.[136,192,193] The reorientation of the active stroke of the beat coincides with a change in the distribution of calcium ions in the superficial cytoplasm.[91] It has been suggested that the calcium ions are involved in some contractile activity which alters the orientation of the ciliary beat. A change in the direction of movement of the metachronal waves follows immediately after a change in beat direction. The reorientation of cilia takes place almost simultaneously over the whole surface of a ciliated protozoon, and Y. Naitoh and R. Eckert[182] have provided confirmation of the hypothesis that the co-ordination of beat direction depends upon membrane potential, depolarization of the membrane leading to reversed beating and hyperpolarization to enhanced forward beating. There is now substantial evidence that neither the co-ordination of reversed beating (e.g. *Opalina, Paramecium, Euplotes*) nor ciliary metachronism are dependent upon fibre systems associated with the ciliary bases.[72,183]

The relationship between ionic distribution and ciliary reversal is believed to be the basis of the phenomenon of galvanotaxis in ciliated Protozoa, and the evidence for this has been discussed by T. L. Jahn.[122] In an electric field the cilia at the cathodal side of the body reverse their direction of beating, the area of the body surface affected being greater at higher current intensities. This reversal is explained as resulting from a reduction in the concentration of divalent calcium ions in relation to monovalent cations at the cathodal side of the organism. In the case of *Paramecium*, the reversal of cilia at the cathodal side of the body results in swimming towards the cathode. Other ciliates may adopt other orientations, e.g. oblique galvanotaxis in *Stylonychia*[70] and transverse orientation in *Spirostomum*.

The rate and direction of swimming of flagellate and ciliate Protozoa is influenced by light.[98] *Euglena* is a well known example of a flagellate showing positive *phototaxis*. Near the base of the locomotor flagellum is a swelling which is believed to be light sensitive, and the manner in which light falls on this swelling is held to determine flagellar activity. The swimming response is photopositive when the flagellar swelling is periodically shaded by the adjacent red-pigmented stigma, but photonegative when the flagellar swelling is continuously illuminated.

THE NUCLEUS

Structure of protozoan nuclei

All Protozoa have at least one nucleus, and in many species several or even many nuclei may be enclosed within the same body of cytoplasm by a single plasma membrane. Ciliates characteristically have two types of nuclei (macronuclei and micronuclei, p. 183) and at a certain stage in the life cycle of some foraminiferans two types of nuclei are known to be present within the same organism (p. 85); the special features of these forms are considered later. All nuclei contain the genetic material deoxyribonucleic acid (DNA) in the chromosomes within the nucleoplasm. The form of nuclei and their manner of multiplication are more diverse in Protozoa than in higher animals or plants. Most commonly protozoan nuclei are vesicular and spherical or oval, the best-known exceptions being the macronuclei of ciliates which are more dense and often have complex shapes.

The **nuclear envelope** consists of two unit membranes about 7 nm thick separated by a space of 20 nm or so, as in other eucaryote cells.[198, 247] The inner and outer membranes of the envelope come together to form the margins of pores which may allow continuity between the cytoplasm and the interior of the nucleus (Fig. 2.18). These pores are commonly 50 nm or

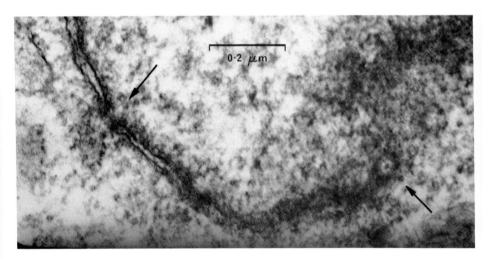

0·2 μm

Fig. 2.18 Electron micrograph of a section of the sporozoan *Selenidium terebellae* showing part of the nuclear membrane with nuclear pores seen in section (at the left) and in surface view (at the right). Micrograph by A. G. H. Dorey (unpublished).

more across, and appear in surface view as dense rings with a number of darker granules embedded in them. In sections through the envelope a septum or zone of denser material is frequently seen across the opening of

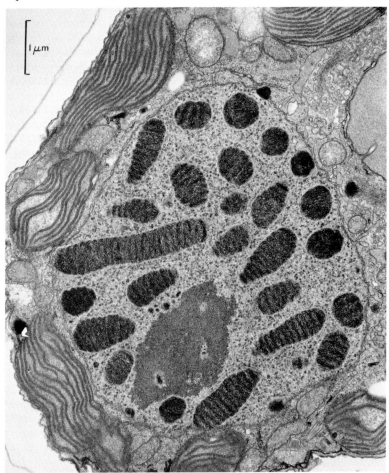

Fig. 2.19 Electron micrograph of a section of the dinoflagellate *Heterocapsa triquetra* showing the appearance of interphase chromosomes in members of this group; note the fine coiled filaments within the chromosomes. Micrograph by J. D. Dodge.[60a]

the pore, so the pores may not represent sites of completely free communication between the nucleus and the cytoplasm. The outer membrane of the nuclear envelope is frequently seen to communicate with elements of the endoplasmic reticulum, and in some flagellates it has been found that the outer membrane of the nuclear envelope is continuous with the outer membrane of the plastid.[86]

A curious honeycomb structure is associated with the inside or outside of the nuclear membrane in some species. In *Amoeba proteus* this consists of a layer of hexagonal membranous tubes about 140 nm across and 300 nm long extending into the nucleus from the inner membrane, where each tube surrounds a pore in the envelope. In other species the pattern is generally less well developed and there is scanty evidence for the function of the honeycomb zone.

In electron micrographs the interior of the interphase (non-dividing) nucleus has a characteristic finely granular appearance, with occasional aggregations of denser material which may represent **chromatin** (DNA) **bodies** (Figs. 2.4, 2.21). Helically coiled fibrils have been seen in the nucleus of *Amoeba proteus*, but these are not present during mitosis when a mass of dense chromatin material appears. In the nuclei of some Protozoa, notably some dinoflagellates[58] and hypermastigote flagellates, the chromosomes are visible throughout interphase in the configuration of chromosomes during mitotic prophase (Fig. 2.19). Protozoa of some groups have a single (haploid) set of chromosomes, while in nuclei of other groups a double (diploid) set of chromosomes is present. It is believed that the nuclei of a number of Protozoa are polyploid, with many chromosome sets in each nucleus. In ciliates the micronucleus is diploid and the macronucleus is normally polyploid (see p. 183).

Nuclei usually contain one, several or many bodies called **nucleoli** associated with one or more of the chromosomes. These bodies are rich in RNA, principally as ribonucleoprotein, but lack DNA; they are associated with the synthesis of ribosomes, whose main components are made within the nucleus, but not finally joined together until they have passed through the pores in the nuclear envelope and entered the cytoplasm. The large nucleolar bodies which occur at the centre of some vesicular nuclei of Protozoa are often called karyosomes, and in these cases the chromatin material is found around the periphery of the nucleus. Nucleoli disappear during nuclear division, but similar RNA-containing bodies called endosomes persist intact during mitosis. Ciliate micronuclei generally lack nucleoli.

The nucleus during the cell cycle[205]

The nucleus has two major functions, the replication of the genetic material of the cell and the release of genetic information to the synthetic machinery of the cell. The genetic information content of the nucleus encoded in the base sequence of the chromosomal DNA is involved in both of these functions. Replication involves the synthesis of new DNA to duplicate the chromosomes, and subsequently the separation of chromosomes into daughter nuclei, normally immediately before cell division. The synthesis of DNA in a nucleus normally occurs at a characteristic time in the cell cycle, and this is usually at a different time in the cycle from the division of the nucleus. The cell cycle may be divided into 4 sections: the

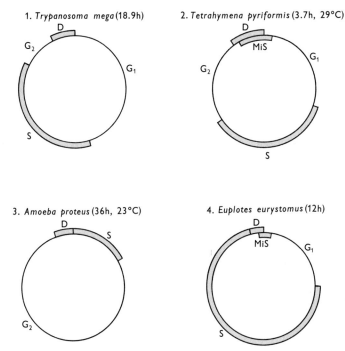

Fig. 2.20   Diagrams to show the proportion of time occupied by the main phases of the cell cycle in four Protozoa. In the ciliates the micronuclear S period (MiS) is shown within the circle for comparison with the timing of events in the macronuclear cycle shown around the outside of the circle. An average figure for the duration of the cycle is indicated after the name of the organism. Information mainly from Prescott and Stone.[205]

division period (D), the time from the end of division to the beginning of DNA synthesis (G_1), the DNA synthesis period (S) and the time from the end of DNA synthesis to the beginning of division (G_2). The timing of these events in some protozoan cycles is shown in Fig. 2.20. In some ciliates the S period for the micronuclei and macronuclei has been found to occur at different parts of the cycle.

There is evidence in some cells that once the transition from G_1 to S has taken place the events leading up to nuclear division and the subsequent cell division have been set in motion. G_1 may thus be the most variable phase. D. M. Prescott and G. E. Stone[205] found that if *Tetrahymena* is deprived of the amino acids tryptophan and histidine during G_1, then there is only partial DNA synthesis and no cell division; if the animals are deprived of these amino acids after the G_1–S transition, then DNA synthesis is completed and cell division follows. It is concluded that before the beginning of DNA synthesis these amino acids are required for the synthesis of the enzymes thymidine synthetase and thymidine kinase, at least one of which is required for synthesis of thymidine triphosphate during DNA synthesis. It is assumed that these enzymes are broken down after DNA synthesis and made anew before each S period, and if the amino acids are not available at the appropriate moment then the enzymes are not synthesized and DNA synthesis cannot take place.

Simultaneous with the synthesis of new DNA there is active synthesis of nuclear histone proteins, which also double in quantity. This has been

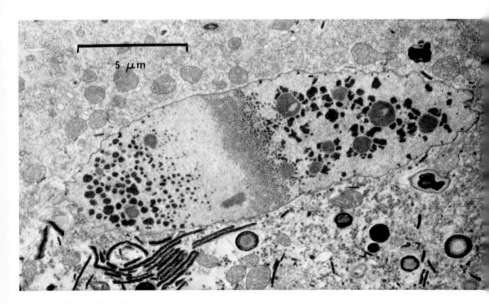

Fig. 2.21 Electron micrograph of a section through *Euplotes eurystomus* showing part of the macronucleus during the passage of a 'replication band' from left to right. Both before synthesis of DNA (at the right) and after synthesis (at the left) the nucleus contains smaller, more dense chromatin masses and larger, less dense nucleoli; both of these structures disappear during the passage of the replication band. Micrograph by R. Gliddon (unpublished).

studied in the macronucleus of *Euplotes*, in which, as in many other ciliates, the new synthesis of DNA occurs as 'replication bands' travel along the nucleus (Fig. 2.21). In *Euplotes* zones of reorganization are seen to originate at either end of the elongate nucleus and to move towards the centre. The passage of a replication band over a section of the nucleus is associated with a doubling in the quantity of both DNA and histones. It is interesting that the replication bands indicate that synthesis of DNA begins simultaneously at both ends of the macronucleus. The signal for the commencement of macronuclear DNA synthesis is not known—it appears to be specific, for the micronuclear DNA does not replicate at the same stage in the cell cycle, but during and immediately after the micronuclear telophase (Fig. 2.20d). Other proteins are also accumulated by the nucleus; in *Amoeba* they may be released on the breakdown of the nuclear membrane during mitosis, but are reaccumulated within 30 minutes or so after re-formation of the nuclear envelope. The importance of these types of protein in nuclear functioning is not established, but many of them must be enzymes and many are believed to be concerned in the mechanisms controlling the release of genetic information to the cell.

Nuclear division

Division of the protozoan cell is usually immediately preceded by division of the nucleus, except in some multinucleate forms where it may be possible for the organism to separate into nucleated parts without an associated phase of nuclear division. The genetic function of the nucleus demands that at division each daughter nucleus should receive a complete set of chromosomes identical with that possessed by the parent cell; this is achieved by the complex processes associated with mitosis (karyokinesis). There have been many reports that the nuclear division in certain Protozoa is not mitosis of the conventional type, but most of these reports have subsequently been found to be erroneous and the process normally differs only in minor ways from mitosis in cells of higher plants and animals. Haploid and diploid nuclei undergo a mitotic process in which each chromosome divides longitudinally into two identical parts, the replicate chromosomes presumably resulting from DNA synthesis in the preceding S period; then the twin products of each chromosome division move apart so that identical groups of chromosomes come together in each of the two daughter nuclei. The situation in polyploid nuclei is less certain, but it is thought that normally several complete chromosome sets (genomes) separate into daughter nuclei and that a balance between chromosomes of various types is maintained. Such nuclei usually undergo changes in shape before their constriction and division, and in some cases non-mitotic chromosome splitting is known to occur before division.

The first sign of an approaching nuclear division is usually the multiplication of the centriole, or its equivalent, and the migration of daughter centrioles to opposite sides of the nucleus (Fig. 2.22). The *mitotic spindle* of microtubular fibrils forms between the centrioles and is important in later stages of the process of mitosis. Protozoan centrioles have the typical

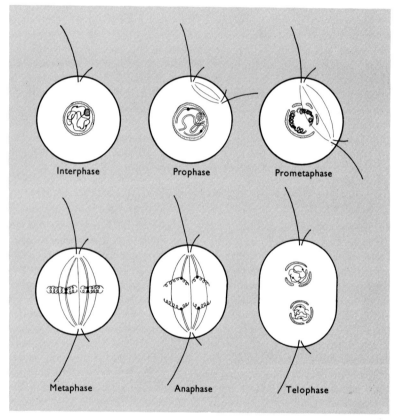

Interphase   Prophase   Prometaphase

Metaphase   Anaphase   Telophase

Fig. 2.22 Stages in the mitosis of a flagellate protozoon. These stages are described in the text.

structure which is also found in flagellar basal bodies; indeed in some flagellates the basal bodies of active flagella act simultaneously as the centrioles for the mitotic spindle. Around the centriole there is often an area of cytoplasm which stains deeply with ferric haemotoxylin and which is referred to as the centrosome. Centrioles are not involved in nuclear division in ciliates or in Protozoa of some other groups, and in many amoebae the centrosome region is replaced by a larger and more diffuse centrosphere. A mitotic spindle is usually found in these forms, even where the centriolar organizing centre is absent.

Before the beginning of mitosis the nucleus contains very long filamentous *chromosomes* and clear nucleoli within the nuclear envelope (Fig. 2.22). The chromosomes progressively shorten and thicken during the prophase of mitosis, and when they are shortened they may each be seen to be double—each chromosome consists of two chromonemes (or chromatids). The chromosomes at this time may appear banded or beaded with aggregations of DNA, and also are constricted at the centromere (kinetochore). During the metaphase stage the centromeres form attachments with fibrils of the spindle near its equatorial plane, and typically become arranged across the equator of the spindle. By this time the nucleoli and the nuclear envelope have normally disappeared. The centromere of each chromosome splits and the two chromatids are separated as the centromeres move towards the poles of the nuclear spindle during anaphase; the mechanism of this movement is unknown, but may depend on changes in length of spindle fibres, some of which are continuous throughout the whole length of the spindle, while those attached to the centromeres are shorter. In the final telophase stage of mitosis the nuclei are reconstructed, the chromosomes lengthen and the nucleoli and nuclear envelope are reformed.

While the nuclei of many Protozoa undergo a mitotic division which fits this description closely, there are a number of ways in which various species differ. Frequently the nuclear envelope persists during mitosis and the nuclear spindle of microtubular fibrils lies within the envelope. In some cases, e.g. the radiolarian *Collozoum*[105] and during gametogenesis of the foraminiferan *Myxotheca*,[229] it has been proved that centrioles lie just outside the nuclear envelope near the poles of the spindle, the spindle fibres terminating in a dense plate at the inner nuclear membrane. In some cases the spindle is formed across one side of the nucleus and the chromosomes are only attached to it by terminal centromeres, e.g. some Coccidia, and some flagellates;[67] in these cases the normal metaphase distribution of chromosomes is not seen, but the migration of chromosomes in anaphase results in a normal pattern of nuclear reorganization. In ciliate nuclei the envelope persists but centrioles have not been seen; the micronuclear mitosis involves an intranuclear fibrous spindle[262] (Fig. 2.23), but macronuclear division is very variable.[209] Normally it is not possible to see any traces of the normal mitosis in the division of polyploid ciliate macronuclei; the large, sometimes multiple, macronuclei usually condense to a single compact body which extends, constricts and separates in a process of 'amitosis'. In *Nassula* the constriction of the macronucleus is preceded by a splitting of shortened chromosomes without participation of a spindle of fibres (endomitosis), although J. B. Tucker has described an intranuclear bundle of microtubules which appears to push the daughter nuclei apart.[262] In a number of cases macronuclear division is accompanied by the elimination of some DNA material into the cytoplasm (hemixis); reabsorption of this DNA occurs later. The significance of this phenomenon of hemixis, which occurs regularly in some species, is unknown.

The mitosis of some phytoflagellates has unusual features. In the

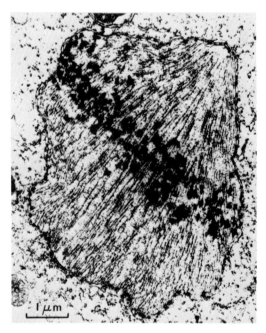

Fig. 2.23 Electron micrograph of a section through a micronucleus of the ciliate *Nassula* at the metaphase stage of mitosis. Micrograph by J. B. Tucker.[262]

euglenids the nuclear membrane persists throughout mitosis and surrounds an internal spindle of microtubules, but the chromosomes lack centromeres and do not all move together during anaphase.[154] There is only a temporary association between the nucleus and the flagellar bases early in mitosis. In dinoflagellates the nuclear membrane again remains intact, but the nucleus is pierced by membrane-enclosed cytoplasmic channels, through which pass groups of microtubules;[63, 149] these microtubules do not make direct contact with either chromosomes or flagellar basal bodies.

The dinoflagellate 'nucleolus' and the euglenid endosome both persist throughout mitosis, during which these bodies divide.

Nucleic acids and protein synthesis

The genetic information of the nucleus is expressed in the synthesis of proteins, both structural molecules and enzymes. While the synthesis of

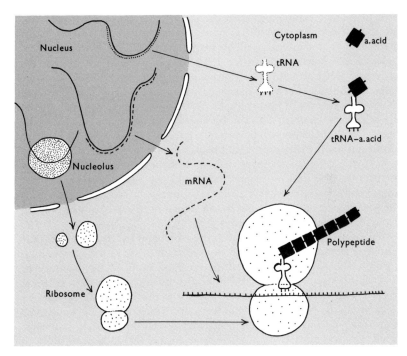

Fig. 2.24 Diagram to illustrate the sequence of events during protein synthesis, as described in the text. m RNA is messenger RNA, t RNA is transfer RNA and a acid is an amino acid.

proteins occurs principally in the cytoplasm, it is dependent upon the continued presence of the nucleus, for protein synthesis ceases soon after the removal of the nucleus. It is now known that the ribosomes are the sites of protein synthesis, and that the protein which will be synthesized is specified by a short-lived RNA molecule called messenger RNA which becomes temporarily associated with the ribosome (Fig. 2.24). The messenger RNA carries a transcript of the DNA genetic code in the form of triplets of nucleotide bases whose linear sequence is determined by base pairing during the synthesis of the messenger RNA against a DNA strand in the nucleus. A ribosome attaches to one end of the messenger RNA molecule and passes along to the other end. As the ribosome passes each triplet of bases of the messenger a temporary linkage is made by base pairing with a molecule of transfer RNA coupled to a particular amino acid of the type specified by the triplet of bases on the messenger. There is a species of transfer RNA for every triplet of bases that occurs in messenger RNA as a code for an amino acid. In the cytoplasm the transfer RNAs form temporary compounds with the appropriate amino acid under the influence of specific activating enzymes, so that a pool of transfer RNA-amino acid compounds is available in the cytoplasm for use at the ribosomes. Each successive transfer RNA which is lined up against the messenger RNA strand at the ribosome brings its amino acid into close contact with the amino acids specified by the previous triplets and brought by previous transfer RNA molecules; this contact leads to the formation of a peptide bond which adds the amino acid to a linear polypeptide chain. The polypeptide is completed when all triplets of the messenger RNA have been 'read' by the ribosome, and it may then fold up in a specific manner to form the complete protein, with or without combination with other polypeptides. Several ribosomes may read the same messenger RNA strand, and it is believed that a ribosome may associate successively with different messenger strands and be concerned in the synthesis of proteins of different types.

While the synthesis of proteins at the ribosomes in the cytoplasm only consumes amino acids and energy derived from stores in the cytoplasm, it requires RNA of three types which must be supplied by the nucleus. Radioactive tracer studies show that the nuclear RNA first becomes labelled with radioactivity and that later the labelled RNA appears in the cytoplasm; in *Tetrahymena* RNA formation ceased in enucleated fragments, while active RNA synthesis continued in nucleated fragments (RNA comes from the macronucleus in ciliates[205]). Ribosomal RNA is needed for synthesis of all proteins and originates from the nucleolus. Transfer RNA of all types is required continuously in active cells and is also of nuclear origin. Messenger RNA strands are believed to be produced only when required for the production of specific proteins at particular times.

Many proteins, including numerous cell enzymes, must be continuously required by the cell, and the messenger RNA for such proteins is probably continuously produced. Marked changes in the production of other enzyme proteins occur during the life of a protozoan cell, e.g. during the DNA synthesis phase, and during the phase of division when new cell organelles are being constructed; the formation of each of these special proteins results from the induced synthesis of the appropriate messenger

RNA as a result of the receipt of some chemical signal. Cells which show more complexity in their life cycle, such as the development of a spore or cyst stage, show many characteristics of the differentiation of cells of multicellular organisms. In the soil amoeba *Acanthamoeba*, for example, the encystment process has been shown to involve new macromolecular syntheses which result in the production of first an outer protein coat and later an inner cellulose layer.[184] The formation of the cyst wall only occurs in the presence of the nucleus; it involves the synthesis of new RNA and new protein, and is blocked by inhibitors of these syntheses. Selective transcription of the genetic information content of the nucleus occurs in Protozoa just as it does in other organisms, but is less extensive than in highly differentiated animals and plants where the characteristics of a large diversity of cell types within an individual are specified by the same set of chromosomes.

EXTRACELLULAR PRODUCTS

Materials manufactured within protozoan cells may be passed to the outside of the cell membrane for some function in relation to the life of the organism. The best-known forms of this secreted material are permanent or temporary shells, tests or cyst walls. Many Protozoa produce quantities of mucus, apparently as an aid in locomotion or feeding, and the production of such extracellular structures as trichocysts and similar threads is widespread. Waste products of metabolism, which may move by diffusion or with fluid expelled by the contractile vacuole, and undigested remains released from spent food vacuoles are also passed out of the cell, but are not included under this heading.

The pellicles of ciliates and of such flagellates as *Euglena*, and the pseudopods of some amoebae, may be strengthened or stiffened with protein fibres or skeletal rods or plates of calcareous or siliceous material, but these structures are within the cell membrane; by contrast, *Chlamydomonas* and related flagellates have a 'cellulose' cell wall outside the membrane comparable with that found in many higher plants. On the outer surface of the membrane of many flagellates and some amoebae are adherent scales or spines which are also clearly extracellular products. Incomplete envelopes found around various Protozoa are variously described as shells (or tests) when they fit closely, e.g. around some amoebae (p. 166), or as loricas when they are chambers within which flagellates or ciliates move more freely (pp. 122 and 208). These envelopes are often composed of a proteinaceous or mucopolysaccharide secretion, usually described as 'chitin', 'pseudochitin' or 'tectin'. This may compose the entire structure of the shell of the amoeba *Arcella* (p. 166) or the lorica of the peritrich ciliate *Cothurnia* (p. 208), or it may be reinforced by foreign materials, such as sand grains or diatom frustules, in testaceous amoebae like *Difflugia*, tintinnid ciliates like *Codonella* (p. 213) or foraminiferans like *Haliphysema* (p. 174). Further secreted material, in the form of internally prefabricated siliceous scales, is incorporated in the shell of the testaceous amoeba *Euglypha* (p. 166), and in most foraminiferans the organic layer is supported by a heavy secretion of calcareous material. Gelatinous or mucous secretions frequently aggregate debris from the environment to form a temporary loose lorica around flagellates or ciliates, and may also provide a means of temporary or more permanent colony formation. Secreted loricas may be stalked, and in some cases naked Protozoa may be attached by secreted stalks or other means of anchorage; the extracellular sheath of the stalk of such peritrich ciliates as *Vorticella* (p. 206) is composed of proteinaceous material of the keratin type.

Many Protozoa form resistant spore or cyst walls at certain times in the life cycle, particularly in such parasites as Sporozoa (p. 224), or in response to certain environmental conditions, as discussed in chapter 11. These walls, which are secreted as close-fitting extracellular coats, may involve a variety of materials; e.g. in *Acanthamoeba* a phosphoprotein layer is secreted first, followed by an inner cellulose layer, while in the coccidian *Eimeria* the very resistant wall of the oocyst is formed of an outer layer of quinone-tanned protein and an inner lipo-protein layer.

Materials secreted to form extracellular structures, including scales and spines as well as shells and cyst walls, are known in some cases to appear first in vesicles of the Golgi system before they are carried to the surface and extruded.[171, 172] The production of mucus in membrane-bound mucocysts is probably a comparable phenomenon. Such materials may be released simultaneously over a large area of the cell surface, and may form shells or cyst walls of a characteristic shape as they are moulded around the whole cell or around an extension of the body, e.g. in the formation of a new chamber of the shell of some foraminiferans. Alternatively the secretion may be released more gradually, as in the case of the flagellate *Poteriodendron*, where the cup-shaped lorica is built up from the base with secreted material applied to the edge of the cup as a continuous strand by a prolonged rotation of the body. The sheath of the peritrich stalk is produced from an aboral tuft of cilia called the scopula, but it is not clear how the ciliary structures are transformed into stalk structures.

3

Nutrition, Metabolism and Growth

NUTRITION OF PROTOZOA[97]

All organisms require a supply of materials to build their body substance and a supply of energy for the performance of synthetic reactions and such necessary activities as movement and transport. The necessary materials must provide a metabolic pool of smaller organic molecules which can be used as 'building blocks' for growth or the laying down of food reserves, and as a source of chemical energy to be made available by the process of respiration (Fig. 3.1). While the immediate source of energy for cell processes is normally the breakdown of the smaller organic molecules, the intake of energy into the cell may be from either or both of two main sources. Many organisms obtain energy from the breakdown of organic

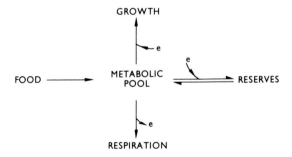

Fig. 3.1 Diagram showing the central position of the metabolic pool of organic molecules in relation to the pathways of metabolism and the utilization and release of energy (e).

materials derived from other organisms, and these organic materials also provide many of the small molecules of the metabolic pool; such organisms practise heterotrophic nutrition. Some organisms practise autotrophic nutrition in which energy derived from light (in photoautotrophs) or from the oxidation of inorganic substances (in chemoautotrophs) is used to build up organic molecules from inorganic materials. Photoautotropic nutrition is found in those flagellates which possess chlorophyll pigments, but chemoautotrophic nutrition is not known to occur in Protozoa—the important chemoautotrophs are bacteria. Many flagellates may use both autotrophic and heterotrophic nutrition, but the nutrition of all other Protozoa is completely heterotrophic.

The organic molecules of which an organism is composed contain a number of important elements which must be present in their intake of nutrients. Carbohydrates contain carbon, hydrogen and oxygen, proteins contain also nitrogen and often sulphur and lipids often contain phosphorus. The nucleotides which form the nucleic acids and such vital compounds as ATP contain nitrogen in purine and pyrimidine bases and phosphorus. An autotrophic organism may be able to synthesize all the complex molecules it requires from simple inorganic compounds containing these elements, together with a few traces of such elements as copper, iron and magnesium. A heterotroph which feeds on the tissues of other animals or plants is likely to take in complete complex molecules of many of the types used in metabolism, and its synthetic powers are often limited. Such a heterotroph may require not only an organic energy source, but also one or more classes of organic molecules as organic carbon sources and organic nitrogen compounds; it may have limited ability to interconvert amino acids, so that it may require many or most of about 20 amino acids as well as a source of purine and pyrimidine compounds. Many organisms, including some that are autotrophs, require a range of growth factors (vitamins) which are generally complex molecules necessary for coenzyme functions.

Nutritional requirements

The nutritional requirements of a few representative Protozoa are shown in Table 3.1. Such flagellates as *Chlamydomonas moewusii* are obligate autotrophs, requiring the intake only of inorganic materials and light energy for survival and growth. This complete dependence on autotrophic nutrition is probably rather rare, for many organisms which can perform photosynthesis are known also to be able to grow heterotrophically in the dark, e.g. *Euglena gracilis*. The forms of organic molecules which can be used, and the range of growth factors required vary widely in different examples; even different strains of *Euglena gracilis* differ in that some will grow on carbohydrate and others will not—all will grow on acetate. Many of these forms have few organic needs, even for heterotrophic nutrition, and have extensive synthetic ability, for which an intake of inorganic nutrients is still necessary. While numerous species can survive adequately

Table 3.1   *Some examples of Nutritional needs in the Protozoa.*
Data mainly from R. P. Hall.[97]

	Group	Energy source	Carbon source	Nitrogen source	Other known needs
Chlamydomonas moewusii	Chlorophyceae (green)	Light	CO_2	NO_3^- (NH_4^+)	None
Polytoma uvella	Chlorophyceae (colourless)	Organic	Acetate (some organic acids)	NH_4^+ (NO_3^- ?)	None
Euglena gracilis	Euglenophyceae (green)	Light or organic	CO_2, Acetate Ethanol (not carbohydrate)	NH_4^+ (NO_3^- NO_2^- amino acids)	Vitamin B_{12} Thiamine
Ochromonas malhamensis	Chrysophyceae (pigmented)	Light plus organic	CO_2, Starch Saccharides Glycerol	NH_4^+	Thiamine Vitamin B_{12} Biotin
Chilomonas paramecium	Cryptophyceae (colourless)	Organic	CO_2, Acetate, Lactate, Ethanol, some fatty acids (not carbohydrate or amino acids)	NH_4^+ (amino acids amides)	Thiamine
Crithidia fasciculata	Kinetoplastida (flagellate, insect parasite)	Organic	Carbohydrate	10 amino acids Adenine or guanine	Thiamine Riboflavin Pyridoxine Pantothenate Biopterin Folic acid Biotin Nicotinic acid
Tetrahymena pyriformis	Ciliata	Organic	CO_2 Acetate Lactate Carbohydrate Amino acids	10 amino acids Guanine Uracil or cytidine	Thiamine Riboflavin Pyridoxine Pantothenate Folic acid Biotin Nicotinic acid Thioctic acid
Acanthamoeba castellanii	Sarcodina	Organic	Carbohydrate Glycerol Acetate, Lactate Amino acids	?	Thiamine Vitamin B_{12}

on either autotrophic or heterotrophic nutrition, it has been shown that *Ochromonas malhamensis* is an example of those flagellates which practise both forms of nutrition and are unable to fix enough energy by photosynthesis for growth to occur; their autotrophic nutrition must always be supplemented by a heterotrophic source of energy. The range of organic molecules required by heterotrophic forms varies widely, particularly as regards amino acids, purines, pyrimidines and growth factors. Generally the flagellates related to autotrophic forms are less demanding than amoebae, and these in turn are less demanding than ciliates and parasitic flagellates.

It is only possible to assess the nutritional requirements of a species if it can be grown in pure (axenic) culture in the absence of any other living organism. It is normally easier to maintain axenic cultures of autotrophic species than heterotrophs, but axenic cultures of Protozoa from many groups have been established and have proved most valuable, not only in nutrition studies but also in research on many other aspects of the physiology and biochemistry of Protozoa. Several species of Protozoa have been used in bioassay procedures for the detection and measurement of very small quantities of vitamins and growth factors;[97] axenic cultures are essential for this work, and precisely defined culture media with high-purity components must be used. Two Protozoa used in this way are *Euglena gracilis* and *Ochromonas malhamensis* in the estimation of vitamin B_{12} at concentrations down to about 1 ng/litre. The Protozoa will grow in culture medium containing the vitamin, but not in the same medium in the absence of the vitamin; the concentration of vitamin in an unknown sample may be estimated by the growth of the protozoon in cultures containing various dilutions of the unknown samples.

BIOCHEMICAL PATHWAYS OF METABOLISM[52]

Studies on axenic cultures have been important in the elucidation and confirmation of biochemical pathways used in various Protozoa for the interconversions of organic molecules to provide energy and the types of molecules required for synthetic reactions. The interrelations of some of the more important biochemical pathways are summarized in Fig. 3.2.

The catabolic (breakdown) reactions shown here are principally concerned with the production of ATP, though the production of reduced coenzymes (especially $NADPH_2$) which may participate directly in synthetic reactions is also important. The catabolism of carbohydrate, fat and protein converges on the tricarboxylic acid (Krebs) cycle, which is coupled to the reduction of coenzymes and ATP production, and which is usually associated with mitochondria. The tricarboxylic acid cycle requires the input of a 2C group from acetyl–CoA (acetyl–coenzyme A complex) to combine with oxalacetate (4C), but several intermediates of the tricarboxylic acid cycle are used as components in synthetic reactions, so that an alternative source of oxalacetate is required. This is provided by the glyoxylate cycle which has been shown to function in some Protozoa.[18] Two acetyl–CoA molecules are used in the production of each molecule of oxalacetate in this cycle, the enzymes for which are found in peroxysomes (glyoxysomes). The glyoxylate cycle is particularly important in Protozoa which use acetate, ethanol or fatty acids as energy sources (and often as carbon sources); some of these forms can assimilate CO_2 into organic molecules by synthetic reactions using energy derived from catabolism. Thus some euglenoid, chlorophycean and cryptomonad flagellates use photosynthesis and heterotrophic oxidative assimilation as interchangeable and essentially equivalent sources of larger organic molecules.

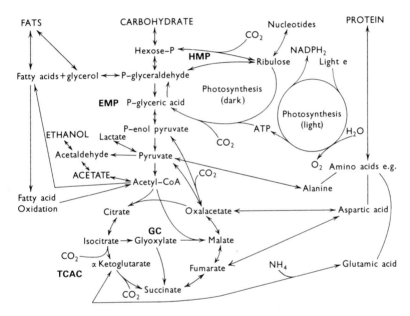

Fig. 3.2 A brief outline of some of the more important biochemical pathways and their interrelationships. Substances named in capitals are common starting points for nutrition in Protozoa. HMP, hexose mono-phosphate pathway: EMP, Embden–Meyerhof pathway; TCAC, tricarboxylic acid cycle and GC, glyoxylate cycle.

The breakdown of carbohydrate by the Embden–Meyerhoff (glycolysis) pathway is better known than breakdown by the hexose monophosphate pathway, but the latter is important because it provides a source of pentoses for nucleotide synthesis. Most reactions of both of these pathways are reversible, and for those steps which cannot be used reversibly for carbohydrate synthesis there are alternative anabolic by-pass reactions. The photosynthesis reactions impinge on these two pathways—the absorption of light energy in the light reactions results in the formation of reduced coenzymes and the production of ATP which are used in the assimilation of CO_2 in the dark reaction cycle of photosynthesis. The synthesis of fatty acids from acetyl–CoA involves a different route from the catabolic process of fatty acid oxidation. Nitrogen compounds are required for the formation of some of the amino acids from pyruvic acid and various organic acids of the tricarboxylic acid cycle; generally NH_4^+ is the most suitable source of nitrogen, although some forms can utilize NO_3^-—no protozoon is known to fix atmospheric nitrogen. These amino acids are used in the production of other acids required for the synthesis of proteins.

The reduced coenzymes ($NADH_2$ and reduced flavin nucleotide) are reoxidized in the mitochondrial cytochrome system (electron transfer chain), with the production of 2 or 3 ∼ (high energy) P bonds and one molecule of water for each pair of H atoms. This oxidation requires molecular oxygen and does not take place in anaerobic organisms where the tricarboxylic acid cycle and oxidation of fatty acids are of no value in energy production; indeed alternative hydrogen acceptors must be found if oxidation reactions are to proceed, so that anaerobic organisms may often produce lactic acid or ethanol. The reduced coenzyme $NADPH_2$ may take part in synthetic reactions or may be used to reduce NAD.

FEEDING

The types of raw materials that must be taken into the body vary widely according to the synthetic abilities of the organism; the needs of photosynthetic flagellates are generally few and less complex than the needs of heterotrophic Protozoa. When the molecules required are small they may enter the body by diffusion or by active transport, and larger molecules are taken in by pinocytosis or phagocytosis (Chapter 2). It is often possible to discriminate between heterotrophic species which take in organic substances dissolved in the surrounding medium without phagocytosis (osmotrophic or saprozoic nutrition) and those which phagocytically ingest solid food originating from animals or plants (phagotrophic or holozoic nutrition); but these two modes of nutrition overlap extensively. The phagotrophic species have the more specialized feeding mechanisms to cope with the capture and intake of the class of food particular to the species; this food might be a single species of animal, protozoon or plant, a wider range of plants or animals, or the protozoon may be an omnivore. A single cytostome with specialized structure is usual in ciliates and phagotrophic flagellates, but most suctorian ciliates and some Sporozoa have many 'mouths' and in amoebae food vacuoles may be formed over a large part of the surface. Parasitic Protozoa may use saprozoic or holozoic nutrition and feed in similar ways to free-living Protozoa, but many Protozoa may be involved in mutually beneficial symbiotic associations, the special nutritional characteristics of which will be mentioned in Chapter 11.

The collection of food by phagotrophic Protozoa may be achieved by the active searching for and capture of food of a particular type, or by various methods of 'fishing' which provide a mixed catch of objects from which particular classes of food particles may be selected; these are roughly equivalent to the macrophagous and microphagous categories of feeding methods employed by animals generally. One common method of microphagous food collection is 'active filtration', involving usually the production of a water current by cilia and flagella and the extraction of

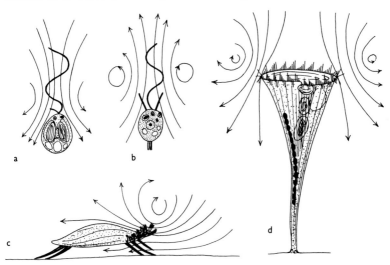

Fig. 3.3 Water currents that bring food to the cell are created by flagella in (a) *Ochromonas*, a member of the Chrysophyceae, and (b) *Codonosiga*, a choanoflagellate, and by the coordinated activity of cilia in (c) *Euplotes* and (d) *Stentor*, which are both ciliates possessing compound feeding cilia called membranelles.

particles from the flowing water. The use of this technique by some flagellates and ciliates is illustrated in Fig. 3.3. The bacteria collected by the flagellar activity of *Ochromonas* are enclosed in food vacuoles by pseudopodial action near the flagellar base, and particles filtered from the water by the collar of *Codonosiga* are taken into food vacuoles which may be seen below the base of the collar.[233] Bacteria are the main food of the ciliates *Vorticella*, *Pleuronema* and *Euplotes*; in these cases the precise role of all of the cilia is not well understood, although it is believed that some cilia are primarily responsible for creating the feeding current while other cilia are concerned with filtration and the passage of food to the cystosome. *Stentor*, *Condylostoma* and many related ciliates are more omnivorous, using their cilia to capture algae, bacteria, ciliates (including other members of the same species), flagellates and even multicellular animals like rotifers.

Another fishing technique is practised by some suctorian ciliates and by many amoeboid Protozoa, in particular among the heliozoans, radiolarians and foraminiferans. In these cases the surface is greatly extended by tentacles or pseudopodia that capture prey organisms which happen to collide with these projections while swimming or floating in the water. The prey adheres to the tentacle or pseudopod and is subsequently eaten. In the case of the suctorians (p. 218), M. Rudzinska[217] has described small bodies, later called haptocysts, which prevent the escape of the prey.

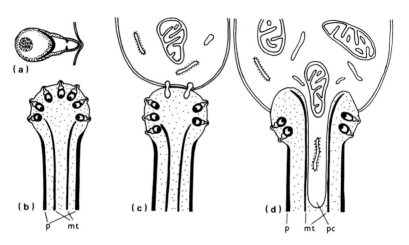

Fig. 3.4 Diagrams to illustrate the appearance (a) and position (b) of haptocysts in a suctorian tentacle, and their role in attachment of the prey to the tentacle tip (c). In (d) the prey cytoplasm (pc) is seen passing down the tentacle within the cylinder of microtubules (mt). The pellicle of the suctorian (p) is less well developed at the tip of the tentacle. Information from micrographs by M. Rudzinska[217] and C. F. Bardele and K. G. Grell.[17]

Haptocysts develop in the body of the ciliate and move up the tentacle to the tip, where the cell membrane is protruded by the underlying organelle (Fig. 3.4). When a prey ciliate makes contact with the tentacle tip, the haptocyst 'discharges' and penetrates the cell membrane of the prey, anchoring it to the tentacle.[17] Subsequently the membrane of the prey is broken down opposite the tentacle tip, and the contents of the body of the prey are sucked into one or more food vacuoles within the suctorian until little more than a part of the membrane of the prey is left and discarded. The tubular tentacle of the suctorian is supported by a cylinder of microtubular fibrils through which the prey cytoplasm enters, but it is not known what part these microtubules may play in the feeding

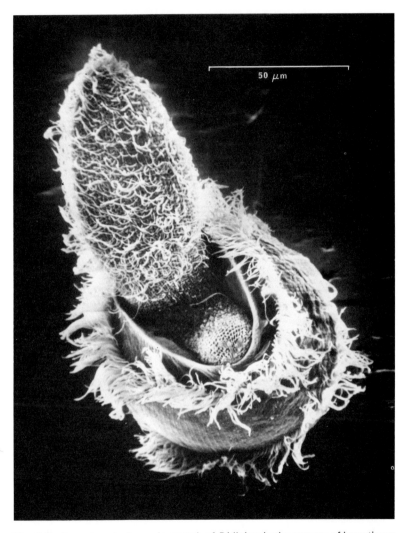

Fig. 3.5 Scanning electron micrograph of *Didinium* in the process of ingesting a *Paramecium*. The proboscis of the predator has been greatly dilated to allow the prey to be drawn in. Compare with a non-feeding *Didinium* in Fig. 8.7a. Micrograph by E. B. Small and D. S. Marszalek.[242]

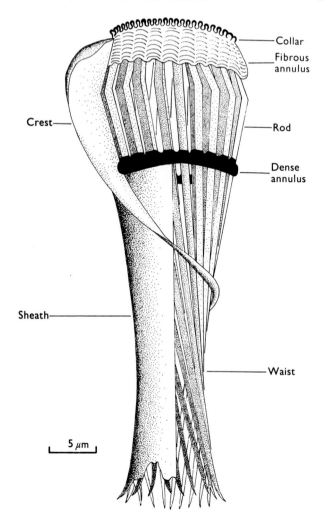

Collar
Fibrous
annulus

Crest—

Rod

Dense
annulus

Sheath—

Waist

5 μm

Fig. 3.6 Diagram showing the structure of the pharyngeal basket of *Nassula*; the collar is covered by the cell membrane and forms the base of a small depression of the body surface, while the lower end of the basket projects deep into the cytoplasm. Illustration by J. B. Tucker.[263]

process. It is generally agreed that suctorians feed by suction, but how the suctorian cell generates the necessary force is not fully understood.[15]

Structures of the haptocyst type have also been found by C. F. Bardele[12] in small centrohelid heliozoans, where it is believed that they are used in prey capture prior to the formation of a food vacuole around the prey by pseudopodial flow. Ciliates and flagellates also appear to adhere to the tentacles of *Actinophrys*, prior to the formation of food vacuoles, but the cause of adhesion here is not certain since haptocysts are not present. Foraminiferans generally seem to prey on smaller, less active organisms which somehow adhere to the fine branches of the pseudopodial network before being enclosed in food vacuoles and carried along in the cytoplasmic flow of the pseudopod.

The feeding mechanisms of those Protozoa which seek out particular forms of food tend to be the most closely adapted. Although many of the gymnostome ciliates tend to be omnivorous, e.g. *Dileptus* (p. 194), which eats detritus as well as animal and plant material, other members of the group are specialized as either carnivores or herbivores. An example is *Didinium*, which eats *Paramecium*; the predator pursues the other ciliate in the water, performing jabbing movements with its proboscis directed forwards, until the proboscis makes contact with a *Paramecium*, discharges many trichocysts and penetrates the prey.[289] Subsequently the fibrous tube of the proboscis dilates and the entire *Paramecium* is taken into the body through the proboscis (Fig. 3.5).

The complex cytopharyngeal basket of cyrtophorine gymnostomes (p. 194) is commonly used for the intake of filamentous algae or diatoms. In *Nassula*,[263] for example, the cylindrical basket is 60 µm long and 15 µm in diameter, and its walls are formed by about 30 longitudinal rods, composed of microtubular fibrils, together with other associated groups of microtubules and filaments (Figs 3.6, 3.7). J. B. Tucker has described changes in the shape of the basket, involving dilation, twisting and constriction of the cylinder, which occur when the cytosome of this ciliate makes contact with a filament of the blue-green alga *Phormidium* and during the subsequent entry of the filament into the basket (Fig. 3.8); once one end of the filament, or a bent loop, has passed through the basket, more of the filament is drawn in without further change in shape of the basket. The algal filament is coiled within the body of the ciliate and, if it is too long to be completely ingested, the inward movement of the filament stops and the free end or ends of the filament are broken off by the activity of the ciliate. The cells of the algal filament separate and groups of them are later seen in separate food vacuoles.[263]

Flagellates may practise a similar form of active predation. *Peranema* (Fig. 5.19f, p. 133) is a euglenoid flagellate which may engulf such organisms as *Euglena*, which may be as large as itself. The cytostome lies beside the flagellar invagination of *Peranema*, and is supported by two curved skeletal rods, whose main structural component is a bundle of microtubular fibrils (Fig. 3.9); the role of these rods and associated structures in the ingestion of food is not completely clear, but they may be protruded and withdrawn for food breakdown or capture. This flagellate is not a specialized feeder, for it has been found to consume detritus, bacteria, flagellates and algae.

Those amoebae that have lobose pseudopodia engulf organisms which

Fig. 3.7 Electron micrograph of a cross section through the dense annulus region of the pharyngeal basket of *Nassula*. The rods lying within the dense annulus are constructed of tightly-packed longitudinal microtubules, while in the crests that extend outside the dense annulus the microtubules are regularly arranged in a more open pattern. The cytoplasm within the basket is less granular than that outside, but contains mitochondria and much vesicular material. Micrograph by J. B. Tucker.[263]

they encounter as they creep around. The smaller amoebae mostly eat bacteria. Larger amoebae such as *Amoeba proteus* and *Chaos carolinensis* are carnivorous forms capturing flagellates, ciliates and even rotifers, but the means of immobilizing these active prey while they are being enclosed by 'anterior' pseudopodia to form a food vacuole remains a mystery.

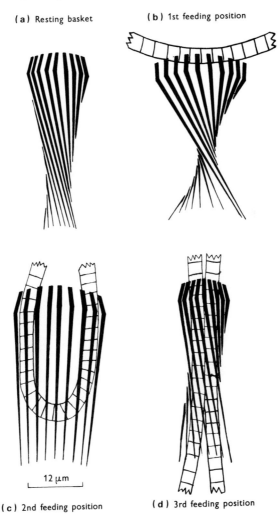

Fig. 3.8 Diagrams showing the changes in shape seen in the pharyngeal basket of *Nassula* during the intake of an algal filament. Illustration by J. B. Tucker.[263]

Pelomyxa palustris is an example of a herbivorous amoeba; it feeds on diatoms and similar algae which are taken in at the tail end of the mono-podial amoeba (see p. 157).

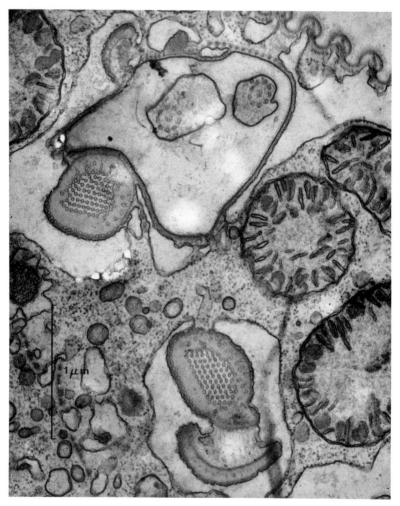

Fig. 3.9 Electron micrograph showing a cross section of the anterior region of the flagellate *Peranema*. The two flagella are seen sectioned within the flagellar invagination (top centre), and below are two groups of microtubules enclosed in a homogeneous matrix that are the pharyngeal rods used in food capture. The mito-chondria have a characteristic type of discoidal cristae. Micrograph by B. Nisbet (unpublished).

For a long time it was assumed that the majority of parasites lived by saprozoic absorption of fluid nutrients, but it has been established recently that in several Sporozoa the protozoon engulfs the cytoplasm of erythocytes into food vacuoles,[43, 218, 219] and it is presumed that this food material from the host cell is taken in through specialized cytostomes described from electron micrographs (p. 224).

GROWTH

In the presence of adequate food most Protozoa go through regular cycles of growth and division (Chapter 4). Since the growth phase usually involves a doubling in volume and the division usually produces only two offspring, the size of the individual organisms in most species does not vary greatly. The characteristics of the growth of such Protozoa in culture may be studied by observing changes in populations living under diverse conditions, and will be discussed in Chapter 11. More extended growth followed by multiple division to produce many offspring occurs in members of most groups, and is especially common in parasites. In these forms the life cycle is usually a complex one involving different stages with specialized morphological features. Growth in all of these forms involves a continuous synthesis of materials for cell structures, although certain organelles and such materials as DNA may only be produced at restricted times during the life cycle; such discontinuous activities are associated with reproduction, which forms the subject of the next chapter.

4

Reproduction and Sex

Increase in numbers in higher animals and many plants is generally associated with sexual processes of a rather stereotyped form. In Protozoa a variety of forms of reproduction is seen,[67] many of which involve multiplication by processes having none of the features of sex, but sexual processes of a wide range of types also occur, and in some cases these processes are not directly correlated with an increase in numbers. The essential features of sexual processes are the differentiation of sex cells (gametes), or at least gamete nuclei, and the fusion of two nuclei at fertilization; a halving of the chromosome number must occur at some time in the life of organisms which show nuclear fusion. An understanding of the role of the various aspects of the sexual process will require some comment on their genetic significance.

NUCLEAR AND GENETIC ASPECTS OF REPRODUCTION

Division of the nucleus in mitosis and meiosis[92]

In the majority of species of Protozoa every individual is capable of reproduction, which may be a complex process, especially in those species with an elaborate organization which must be replicated. The division of one individual to produce two daughters is the commonest mode of reproduction, but in many species, particularly parasitic ones, this binary fission may be replaced by multiple fission. Division of a cell is normally preceded by nuclear division, by either mitosis or meiosis, although in some multinucleate forms fission and nuclear division are not linked, so that for example new individuals may be formed by plasmotomy in which the body is simply separated into multinucleate masses—at any time some nuclei may be found in mitosis in such organisms. In the growth and division cycle of most Protozoa the synthesis of DNA and replication of chromosomes is restricted to a part of the interdivision period which is often separated from the division period by a phase of growth (p. 47). A doubling of the amount of DNA by the synthesis of deoxyribonucleotides normally occurs in each cycle of an organism reproducing by binary fission, but in the rapid series of nuclear divisions that may precede multiple division, or during a succession of fissions without cell growth, such as may occur in the formation of spores or gametes, the situation is more complex and often seems to require the availability of a pool of nucleotides for rapid DNA replication.

Some Protozoa have diploid nuclei, with two chromosomes of each type, such as one finds in the somatic cells of most higher animals, and other Protozoa have haploid nuclei with unpaired chromosomes, such as one finds in the gametes of higher animals and plants; polyploid nuclei with several sets of chromosomes also occur in Protozoa. Diploid nuclei may undergo a process of meiosis to produce haploid nuclei, but more commonly both haploid and diploid nuclei divide by mitosis to produce two daughter nuclei like themselves.

Before the start of mitosis every gene has been replicated, and during the process of mitosis (p. 49) every chromosome divides longitudinally to separate the genes and form two chromatids; following division of the centromeres one chromatid of each chromosome passes to one daughter nucleus and the other chromatid goes to the other daughter nucleus. Meiosis differs in that the centromeres do not divide and the members of each pair of homologous chromosomes in the diploid nucleus are separated, so that one member of each pair goes to one daughter nucleus, the other member of each pair migrates to the other daughter nucleus, and the chromosome number of the nuclei is halved. Most commonly meiosis involves two nuclear divisions, but one-division meiosis occurs in Sporozoa and some complex flagellates. In mitotic division both chromosomes and centromeres divide, and Cleveland[33] has pointed out that in a one-division meiosis division of both chromosomes and centromeres is suppressed, while in two-division meiosis a division of centromeres is suppressed in the first division and a division of chromosomes does not occur in the second division. The three processes are compared diagrammatically in Fig. 4.1. One-division meiosis could be the primitive meiotic process, although in those complex flagellates where it occurs today it surely arose by reduction from a full two-division meiosis.

The complete two-division meiosis is normally a more complex process than mitosis, although it has many comparable features. The two divisions usually occur in quick succession. During the extended prophase of the

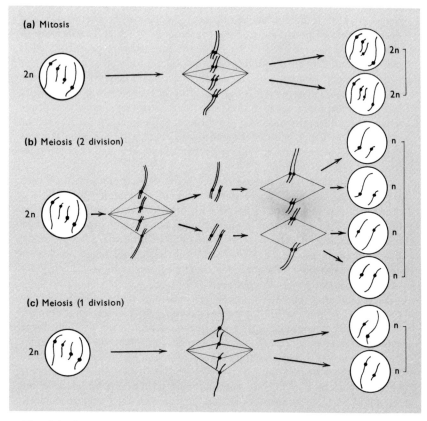

Fig. 4.1   Comparison of the behaviour of chromosomes in (a) mitosis, (b) two-division meiosis and (c) one-division meiosis (see text).

first division the chromosomes divide into chromatids and associate together in homologous pairs (Fig. 4.2), members of a pair becoming so closely coiled together that sections of chromatids may be exchanged between them—a process known as crossing over occurs as a result of chiasma formation between chromatids. Following further shortening of the chromosomes the attachment of the centromeres to the spindle occurs in metaphase, each chromosome having but one attachment since the centromeres are not duplicated here as they are in mitotic metaphase. During the anaphase movement of the first division the two chromosomes separate and crossed-over portions of chromatids separate with the centromeres to which they are now connected, so that two haploid groups of chromosomes are formed. Shortly after this each set of chromosomes becomes arranged on a new spindle for metaphase of the second division. This time the centromeres do divide and the two chromatids of each chromosome separate in the subsequent anaphase to form two haploid nuclei. The final telophase of such a two-division meiosis therefore results in the formation of four haploid nuclei from one original diploid nucleus. Cell division may accompany each nuclear division or may not occur until later. In one-division meiosis the chromosomes do not divide and do not aggregate in close pairs, so that the meiotic prophase is less complex than in two-division meiosis; the opportunity for chiasma formation and crossing over is presumed not to occur, and all chromosomes remain un-

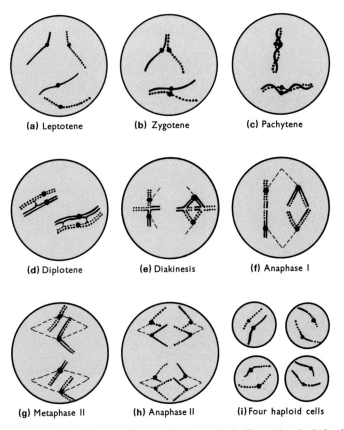

(a) Leptotene   (b) Zygotene   (c) Pachytene

(d) Diplotene   (e) Diakinesis   (f) Anaphase I

(g) Metaphase II   (h) Anaphase II   (i) Four haploid cells

Fig. 4.2   A selection of stages of a two-division meiosis, illustrating the behaviour of chromosomes described in the text.

changed. Only two haploid nuclei will be formed from each diploid nucleus in one-division meiosis.

Patterns of life cycle in Protozoa

Many Protozoa show no trace of any sexual stage in the life cycle; some of these species are believed to have evolved from forms showing meiosis and fertilization, but others may be primitively asexual.[100] The occurrence

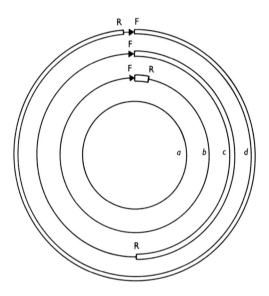

Fig. 4.3 Diagram to illustrate the timing of fertilization (F) and reduction division (R) in four different patterns of protozoan life cycle. In *a* the organism is asexual and haploid, *b* is haploid with zygotic meiosis, *c* is haplo-diploid with intermediary meiosis and *d* is diploid with gametic meiosis.

of meiosis in red algae, which probably had an evolutionary derivation from the eucaryote stock before the appearance of flagella and the earliest Protozoa, suggests either that no Protozoa are primitively asexual or that meiosis evolved independently in red algae and the early flagellates. It is very difficult to be certain that an organism is entirely asexual, e.g. *Amoeba proteus* is one of the most intensively studied protozoans and is almost certainly asexual, but one possible report of sexual reproduction in this species raises an unsatisfied doubt. Species that are primitively asexual might be expected to be haploid, but many asexual species are probably diploid or polyploid—in the absence of meiosis this may be difficult to determine. The life cycle of an asexual haploid protozoan such as *Trypanosoma* is illustrated in Fig. 4.3a. The earliest appearance of sexual processes probably involved the fusion of two haploid individuals, followed by meiosis of the zygote to produce haploid individuals which proceeded to grow and divide mitotically (Fig. 4.3b); such a life cycle is shown by *Chlamydomonas* and some other flagellates and by Sporozoa. In some cases the diploid zygote undergoes considerable growth before division, and an extension of this growth phase with mitotic divisions could give the type of life cycle with intermediary meiosis and an alternation of haploid and diploid phases that is found in some algae, in higher plants and in foraminiferans (Fig. 4.3c). Further extension of the diploid phase leads to the pattern of life cycle that zoologists tend to regard as normal because it occurs in higher animals; diploid organisms which undergo meiosis in the formation of gametic nuclei occur in some groups of flagellates, in the Heliozoa and in the Ciliophora (Fig. 4.3d).

Variants of the sexual process in Protozoa[93]

Sexual processes in Protozoa all require the occurrence of meiosis in a complete or reduced form at some stage in the life cycle, and all involve the production of haploid gametic nuclei (pronuclei), two of which fuse in fertilization to form a syncaryon. After these requirements have been met however, there is still scope for much variation in detail. Gametic nuclei are usually, but not invariably, found in special gamete cells. In some species all gametes look alike (isogametes), but in other Protozoa the gametes may differ in size or structure, so that in many cases it is usual to refer to male and female gametes, by analogy with the anisogametes commonly found in Metazoa. The mating of gamont cells before these have differentiated to form gametes or gametic nuclei occurs in some groups, and the fusion of gametic nuclei occurs after this differentiation; this form of mating is called gamontogamy to distinguish it from gametogamy where two free gametes fuse. Autogamy is the fusion of gametes or gametic nuclei formed from the same gamont. Some of these variants of the sexual process will be illustrated with examples.

Sexual processes in non-ciliate Protozoa

Flagellate members of the Phytomonadida (= Chlorophyceae) are haploid with zygotic meiosis. In some single-celled members of the group, such as certain species of *Chlamydomonas*, normal vegetative cells transform directly into gametes in response to external conditions, while in other species of *Chlamydomonas* gametes are formed after two or more divisions of gamont cells. Gametes produced directly or after division of a phytomonad may be isogametes, or division may result in the formation of macrogametes and microgametes. Normally a microgamete fuses with a macrogamete, but in some species fusion of two macrogametes or two microgametes can also occur. Although isogametes may appear to be similar, there is sexual differentiation at a chemical level and such gametes may normally be allocated to one or other of two 'sexes', referred to as + and −. In dioecious species all members of a clone are of the same sex, and sex is normally determined genetically, but in monoecious species members of one clone may give rise phenotypically to gametes of both sexes. The sexuality of isogametes may be demonstrated in those dioecious species where the phenotype of members of one clone may be modified, e.g. by a change in pigmentation or in the form of food store; when gametes

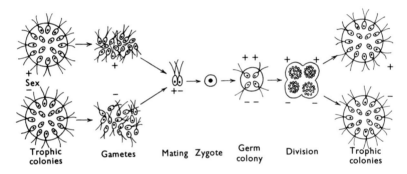

Fig. 4.4   Sexual reproduction in *Gonium pectorale*. Information from K. G. Grell.[93]

of such a clone are mixed with gametes from a normal clone of the opposite sex, every fusion which takes place involves one marked gamete and one normal one. Physiological differences between gametes of the two sexes are also shown by the fact that gametes of one sex may agglutinate on addition of a cell-free filtrate from a suspension of gametes of the other sex, and by the consistent behaviour of gametes of one sex or the other following fusion, when for example the activity of flagella of all members of one sex may cease and each zygote swims using only the two flagella of the other cell of the pair. All cells of some colonial phytomonads develop into free-swimming isogametes, e.g. in the 16-celled colonies of *Gonium pectorale*, where meiosis of the zygote leads to the development of two + and two − clones (Fig. 4.4). In the larger colonial species only a small proportion of cells form gametes, and usually each of these vegetative cells either produces a single large stationary macrogamete or divides to form 16, 32 or 64 motile biflagellate microgametes which are liberated from the colony; some species are dioecious and some monoecious. Among zooflagellates the polymastigote and hypermastigote *metamonad flagellates* studied by L. R. Cleveland show a variety of sexual processes.[34] These flagellates, which inhabit the gut of termites and the wood-eating roach *Cryptocercus*, normally divide asexually, but sexual stages are found in moulting insects and in all cases the occurrence of sexual differentiation of the flagellate is under the control of the moulting hormone (ecdysone) of the host insect. Sexual stages of these flagellates were formerly believed to occur only in moulting nymphs of *Cryptocercus*, but sexual activity of several species has now been reported from termites.

Trichonympha is an example of a haploid hypermastigote flagellate with a zygotic meiosis (Fig. 4.5a). Normal trophic cells transform into gamonts which encyst and lose most of the extranuclear organelles of the cell before dividing to produce two gametes of opposite sexes. New organelles are produced and the gametes emerge from the cyst and swim away. The sex cells come together in pairs when one gamete becomes attached by its anterior rostrum to a posterior fertilization cone of another gamete. The posterior (male) cell has no specialization comparable with the posterior clear fertilization cone and surrounding ring of granules of the anterior cell. The two gametes swim together for a while before the posterior cell is completely drawn into the anterior one, the extranuclear organelles of the male cell are broken down and nuclear fusion occurs. Two meiotic divisions take place within a few hours of fertilization and four cells are produced from the zygote.

The details of the sexual processes vary in other haploid genera. In the polymastigote *Saccinobaculus*, for example, the asexual cell forms a gamont without encystment, and divides to produce two similar gametes (Fig. 4.5b). Gametes fuse laterally, and lateral fusion of the two axostyles (p. 147) from the gametes precedes fusion of the pronuclei; in fact complete nuclear fusion is delayed for some weeks, and the flagella of both gametes are temporarily retained. Following nuclear fusion meiosis takes place in a single division and only two haploid cells are produced from each zygote. Autogamy also occurs in this genus; division of the nucleus of the gamont and replication of its organelles is not followed by cell division, but by fusion of the two axostyles and delayed fusion of the two pronuclei within the same cell to form the zygote.

Macrospironympha is a diploid polymastigote flagellate with a gametic meiosis (Fig. 4.5c). The first meiotic division occurs in a swimming flagellate cell, the two products of which encyst separately before the second meiotic division. The gametes, one male and one female, escape from the cyst and swim away. Fertilization occurs later, the gametes fusing in

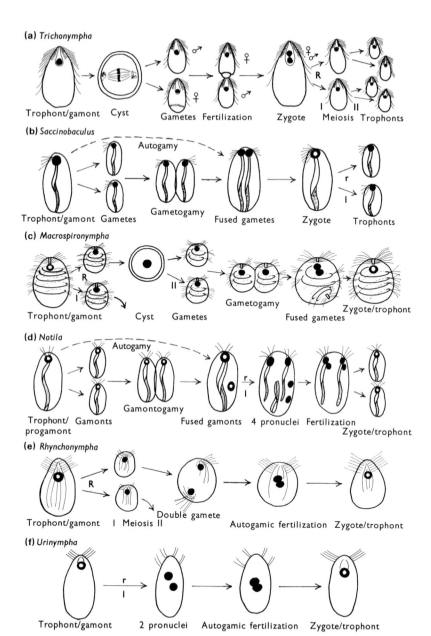

(a) *Trichonympha*

Trophont/gamont   Cyst   Gametes   Fertilization   Zygote   Meiosis   Trophonts

(b) *Saccinobaculus*

Trophont/gamont   Gametes   Gametogamy   Fused gametes   Zygote   Trophonts

Autogamy

(c) *Macrospironympha*

Trophont/gamont   Cyst   Gametes   Gametogamy   Fused gametes   Zygote/trophont

(d) *Notila*

Autogamy

Trophont/   Gamonts   Gamontogamy   Fused gamonts   4 pronuclei   Fertilization
progamont   Zygote/trophont

(e) *Rhynchonympha*

Trophont/gamont   I   Meiosis   II   Double gamete   Autogamic fertilization   Zygote/trophont

(f) *Urinympha*

Trophont/gamont   2 pronuclei   Autogamic fertilization   Zygote/trophont

Fig. 4.5 Diagrams summarizing events during sexual processes of a selection of polymastigote and hypermastigote flagellates, as described in the text; nuclei with white centres are diploid, those that are entirely black are haploid, 2-division meiosis R and 1-division meiosis r. Information from L. R. Cleveland.[34]

almost any position. The extranuclear organelles of the male gamete are lost and those of the female gamete are retained in the zygote, which subsequently becomes a normal asexual individual.

The polymastigote flagellate *Notila* is diploid (or occasionally tetraploid) in the asexual stage and meiosis occurs in the formation of gametic pronuclei (Fig. 4.5d). In the diploid members of this genus two diploid gamonts come together, the two axostyles fuse and one ('male') nucleus separates from its associated organelles. Subsequently each nucleus undergoes a one-division meiosis, the original axostyles are lost and new axostyles develop in association with each of the female pronuclei, while the male pronuclei lie free in the cytoplasm. Each male nucleus fuses with a female pronucleus, and the double zygote formed divides without further nuclear changes to produce two diploid individuals. Autogamy sometimes occurs in *Notila* when a 'progamont' cell, which would normally divide to produce two gamonts, fails to undergo cytoplasmic division following mitosis, and the two diploid nuclei remain within the same cell to take part in meiosis followed by fertilization and the formation of a double zygote.

Autogamy is the only sexual process in the related diploid *Rhynchonympha* (Fig. 4.5e). At the first meiotic division the cell divides, but cell division does not accompany the second nuclear division, and the gametic pronuclei fuse within the same cell. *Urinympha* shows a further reduction in that no cell division takes place and a one-division meiosis produces two pronuclei which fuse autogamously (Fig. 4.5f). Although no changes in genetic constitution occur, there is a replacement of many extranuclear organelles.

The studies of Cleveland on these complex flagellates, which appear to be closely related and live in a similar environment, have revealed a truly surprising variety of cytogenetic life cycles. One can only speculate that diversification or reduction is encouraged in this highly protected environment.

The **Sporozoa** are haploid Protozoa in which multiple mitotic divisions are normally involved in the formation of gametes, and meiosis occurs in the first division of the zygote (Fig. 9.2, p. 225). The cytogenetic events in the sporozoan life cycle are the same as those in *Saccinobaculus* when that flagellate is cross-fertilized. In gregarines two gamonts come together and the pair may remain attached 'in syzygy' for some time and may continue to grow before the formation of a gamontocyst around the pair of cells (Fig. 9.3, p. 227). Within the common cyst each gamont cell undergoes nuclear divisions to form gametes, usually in large numbers; in *Monocystis* these are isogametes, but in some other cases anisogametes occur and one type of gamete may carry one or more flagella. Fertilization takes place within the gamontocyst and each zygote forms a spore within which sporozoites develop. In coccidians some gamonts differentiate into macrogamonts which become macrogametes without division, and others differentiate into microgamonts which divide several times to produce many microgametes that are usually biflagellate (Fig. 9.6, p. 230). The production of free microgametes is usual in coccidians, but in the family Adeleidae a microgamont associates with a macrogamont before differentiation of the microgametes. An oocyst is normally formed around the solitary coccidian oogamete after fertilization, and the zygote undergoes a one-division meiosis, sometimes followed by one or more mitotic divisions; the products of division each form a spore within which sporozoites are formed, except in haemosporidians where the sporozoites are free. Further asexual multiplication involving mitoses and multiple schizogony

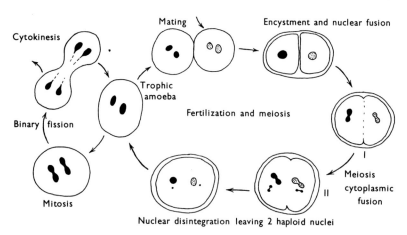

Fig. 4.6 Diagrams of stages in the life cycle of the dicaryotic amoeba *Sappinia diploidea*, showing the behaviour of the nuclei. Information from V. Dogiel.[67]

is common in many sporozoans, and the occurrence of this will be discussed in Chapter 9 when the life cycles of various Sporozoa are considered.

Sexual reproduction occurs in several groups of **amoeboid Protozoa**, and a number of interesting forms of life cycle are found in this group. Information is only included where knowledge is regarded as reasonably complete; for example, flagellate gametes have been reported in Radiolaria, but full details of the life cycle are not available.

Some years ago a curious nuclear cycle was described for the amoeba *Sappinia diploidea* (Fig. 4.6). A normal vegetative amoeba has two nuclei, both of which divide each time binary fission occurs. Sexual fusion involves two ordinary binucleate amoebae which encyst together in a common membrane. The two nuclei within each amoeba fuse, and each resulting single nucleus undergoes a two-division meiosis after which most nuclei disintegrate to leave a single haploid nucleus in each cell. Cytoplasmic fusion of the two cells occurs within the cyst and brings together two nuclei from different amoebae within the same cytoplasm; this new dicaryotic amoeba later emerges from the cyst to take up the life of a vegetative amoeba whose two nuclei will not fuse to form a syncaryon until a sexual phase begins again.[67] Further details of this life cycle and confirmation of its more peculiar aspects would be most welcome. A comparable prolonged dicaryotic stage is common in certain groups of fungi.

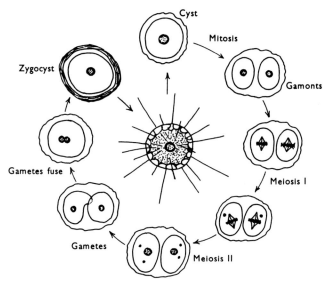

Fig. 4.7 Diagrams illustrating events during the process of autogamy in *Actinophrys* described in the text.[21]

Stimulation of sexual activity in the **heliozoans** *Actinophrys* and *Actinosphaerium* may occur when starvation follows active growth and reproduction. These amoebae are diploid with a gametic meiosis, and autogamy is the normal sexual process. In the mononucleate heliozoan *Actinophrys sol* the sexual process begins with encystment of the amoeba followed by division within the cyst (Fig. 4.7). Each of the amoebae then passes through a two-division meiosis, from which only one haploid nucleus survives, and the resulting cells are gametes. One gamete develops a pseudopodium which moves towards the other gamete to initiate the fusion of the cytoplasm and nuclei of the two gametes, forming a zygote around which a resistant zygocyst wall is secreted; this forms a resting stage which remains dormant until the return of good conditions. The encystment together of two *Actinophrys* within a common membrane has been observed; it results in the formation of two gametes from each cell and fusion of gametes derived from different amoebae may follow, although autogamy is more usual even when two individuals encyst together. The multi-nucleate *Actinosphaerium* divides into a number of uninucleate amoebae before encystment, and thereafter follows the sequence of events described for *Actinophrys*.

Some *foraminiferan* life cycles that have been found to contain sexual phases display an alternation of diploid and haploid generations[93] that is not known to occur in any other Protozoa, but which is familiar because it is found in many plants. The haploid individuals produce gametes by mitosis, and are therefore gamonts; the zygote develops into a diploid

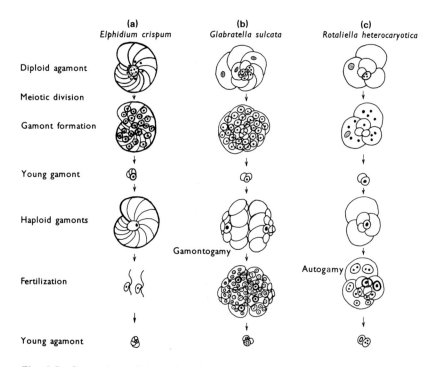

Fig. 4.8 Comparison of three of the life cycle patterns shown by foraminiferans, (a) gametogamy of *Elphidium crispum*, (b) gamontogamy of *Glabratella sulcata* and (c) autogamy of *Rotaliella heterocaryotica*. In (b) and (c) the agamont has larger somatic nuclei and smaller generative nuclei. Information from K. G. Grell.[93]

agamont which undergoes meiosis in the formation of haploid gamonts. There are thus two reproductive phases, separated by intervening growth periods, and meiosis is intermediary. Sexual reproduction may involve gametogamy, gamontogamy or autogamy, and closely related species may show different forms of sexual reproduction.

In the familiar species *Elphidium crispum* (named *Polystomella crispa* in many textbooks) the full-grown gamont produces large numbers of small (2–5 μm) isogametes with two unequal flagella. Fertilization takes place in the sea and the zygote is a small agamont which grows and eventually divides many times with meiosis to form numerous small gamonts (Fig. 4.8a). Such gamontogamous species as *Glabratella sulcata* have a similar life cycle (Fig. 4.8b), but two or sometimes more gamonts come together and gametes unite within the space formed by the two shells; commonly these gametes move by three flagella, and in other cases the gametes are amoeboid. The agamonts are released when the shells separate once more.

Autogamy takes place in *Rotaliella heterocaryotica* and some other species by the formation of amoeboid gametes which fuse in pairs within the shell cavity of the gamont (Fig. 4.8c). Frequently the agamonts of foraminiferans are multinucleate as a result of early mitotic divisions of the zygote, and only some of the nuclei, called generative nuclei, will later provide nuclei for the gamont generation, while the other 'somatic' nuclei disappear early during the division process leading to formation of gamonts. The gamonts are normally uninucleate until gametogenesis begins.

Some other foraminiferans do not show an alternation of gamont and agamont generations. Frequently two or sometimes more generations of agamonts may occur between one gamont generation and the next, and in the case of *Allogromia* it has been reported that these agamonts include multinucleate and uninucleate diploid forms as well as multinucleate and uninucleate haploid forms.

Features of sexual processes in ciliate Protozoa[19, 92, 204]

Ciliate Protozoa generally carry nuclei of two types (p. 183). The macronucleus which is polyploid and has a somatic function disintegrates and is replaced during the sexual process, while the micronucleus is diploid and carries the genetic information through sexual processes. Sexual activity in ciliates commonly involves a special form of gamontogamy known as conjugation, in which each gamont is monoecious and undergoes meiosis in the formation of gametic nuclei; each gamont normally receives a gametic nucleus from the other gamont, and each becomes a zygote. Autogamy is also common. Some variation occurs in the details of these processes, but their main features may be seen in the following example.

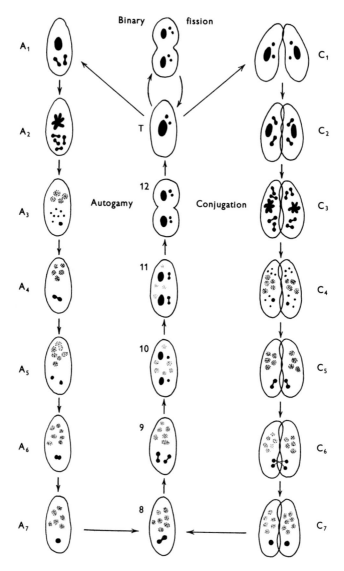

Fig. 4.9 The sequences of events in the sexual processes of conjugation and autogamy in *Paramecium aurelia*; T the trophic ciliate, A_1–A_7 stages in autogamy, C_1–C_7 stages in conjugation and 8–12 stages common to both processes.

Paramecium aurelia has one macronucleus and two micronuclei. At the beginning of **conjugation** two ciliates come together (Fig. 4.9), becoming attached first at deciliated areas near the anterior end of their ventral surfaces, and later by attachments formed in the gullet region where membrane fusion leads to the formation of a number of pores providing cytoplasmic bridges between one gamont and the other. Both micronuclei in each conjugant undergo a two-division meiosis, so that eight haploid nuclei are formed in each cell, seven of which disintegrate. The remaining micronuclei undergo mitosis to produce two gamete nuclei in each cell. One nucleus from each gamont migrates across a cytoplasmic bridge into the other cell and fuses with the nucleus which remained stationary and a syncaryon is formed. Subsequently the cells separate, and by this time the original macronuclei of the two cells have more or less disintegrated. The micronuclear syncaryon divides twice mitotically and two of the four products become macronuclei and two remain as diploid micronuclei. At the first binary fission of the ex-conjugant one macronucleus passes to each daughter cell and both micronuclei divide mitotically, so that the original complement of nuclei is restored.

In **autogamy** a single *Paramecium* undergoes the same sequence of nuclear divisions and disintegrations, but following the mitotic division of the haploid nucleus the two gametic nuclei fuse with each other (Fig. 4.9). When two cells come together as if to conjugate, but end up by performing an autogamous self-fertilization, the process is referred to as selfing or cytogamy. In none of the sexual processes of *Paramecium* does an increase in number of individuals occur, but this may follow by mitosis and binary fission before the normal nuclear constitution is restored.

The pattern of **restoration of the normal nuclear constitution** following conjugation in different species of ciliates is particularly diverse (Fig. 4.10).[92] In a simple case, the syncaryon of *Chilodonella uncinata* divides once and one product becomes the micronucleus and the other the macronucleus. Two divisions of the syncaryon, as found in *P. aurelia*, are common, but the details vary; in *Euplotes patella* two of the products of such a division degenerate and the other two nuclei form one micronucleus and one macronucleus, while in *Didinium* two of the four products fuse to form a single macronucleus and the other two persist as micronuclei. In *Paramecium caudatum* there are three quick divisions of the syncaryon, three of the products degenerate, one becomes a micronucleus and four become macronuclei; the micronucleus divides mitotically at the next two divisions, but the macronuclei merely separate into the four daughter cells and do not divide, so that four caryonides (see p. 93) are formed from each conjugant in this species.

Some ciliates are dioecious, e.g. *Vorticella*, where one gamont contributes only a migratory gamete nucleus and the other contributes only a

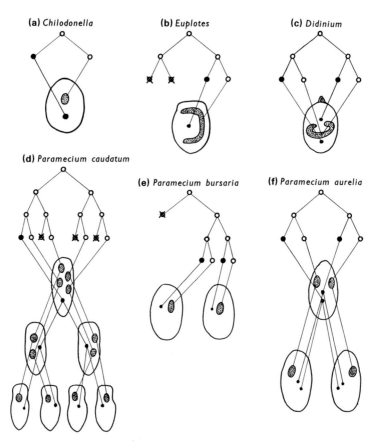

Fig. 4.10 Examples of some patterns of nuclear reorganization following sexual processes in (a) *Chilodonella uncinata*, (b) *Euplotes patella*, (c) *Didinium nasutum*, (d) *Paramecium caudatum*, (e) *Paramecium bursaria* and (f) *Paramecium aurelia*. Information from K. G. Grell.[92]

stationary gamete nucleus. Frequently the cell producing the migratory pronucleus is smaller than the other and is absorbed into the macrogamont during the mating process.

Among the lower holotrich ciliates are forms with only one type of nucleus (*Stephanopogon*) and forms with macronuclei that are diploid (in the families Trachelocercidae, Loxodidae and Geleiidae among others); these diploid macronuclei do not divide, but are always formed directly from micronuclei. Conjugation has not been reported for *Stephanopogon*, but in *Trachelocera phoenicopterus* Raikov found that at conjugation a number of gametic pronuclei are formed in each conjugant; several of these migrate and fuse with stationary pronuclei in the other conjugant to produce several syncarya, but subsequently all of these become pycnotic except one.[209] Other features of the nuclear organization of these forms are considered in the chapter on ciliates (p. 183).

Conditions required for conjugation of ciliates[19, 231, 244]

For some species it is known that two individual ciliates will only conjugate under certain conditions; they must be of the correct mating type, and the age of the clone to which they belong must be within certain limits. Ciliates recognized as belonging to the species *Paramecium aurelia* on morphological characteristics have been separated into 12 varieties. Members of one variety will not normally conjugate with members of another variety, so that each variety is effectively a genetically isolated species (see Table 4.1). Varieties have been called **syngens** by Sonneborn because they represent groups of ciliates which share the same 'gene pool'. Within each syngen of *P. aurelia* there are two **mating types** (except in one syngen where only one mating type has been found), and conjugation will only take place between individuals of different mating type but

Table 4.1 The mating types of *Paramecium aurelia*. Data derived from G. H. Beale.[19] A further 3 syngens are now known.

Syngen	Mating type	Other mating types with which conjugation occurs (and %).	Group	+ or − type
1	I	II (95%), X (40%)	A	−
	II	I (95%), IX (40%), XIII (10%), V (1%)		+
2	III	IV (95%)	B	.
	IV	III (95%)		.
3	V	VI (95%), [XVI (40%)],* II (1%)	A	−
	VI	V (95%)		+
4	VII	VIII (95%), [XVI (95%)]	B	−
	VIII	VII (95%), XV (95%)		+
5	IX	X (95%), II (40%)	A	−
	X	IX (95%), I (40%)		+
6	XI	XII (95%)	B	.
	XII	XI (95%)		.
7	XIII	II (10%)	A	−
8	XV	XVI (95%), VIII (95%)	B	−
	XVI	XV (95%), [VII (95%), V (40%)]		+
9	XVII	XVIII (95%)	A	.
	XVIII	XVII (95%)		.

* Square brackets indicate incomplete mating reactions not resulting in conjugation.

belonging to the same syngen. Mating type specificity in *Paramecium* seems to be expressed in chemical characteristics of the surface membrane, such that contact between mature ciliates of complementary mating types leads to aggregation and attachment in conjugation. The mating type mechanism has no connection with sex, since both conjugants are hermaphrodite; it is better compared with the devices used by flowering plants to ensure cross-pollination.

Following separation of the conjugants of *P. aurelia*, each ex-conjugant gives rise to a clone of cells by successive binary fissions. **Clonal age** influences the ability of individuals to take part in sexual processes; members of a clone are not capable of conjugation until a certain number of fission cycles has been completed since the previous conjugation. Following this period of immaturity, which lasts for 2–10 days, the ciliates are mature for a month or so and will conjugate successfully with ciliates of the complementary mating type. As the clone grows older and enters a period of senescence, fertility decreases, and as a result conjugations become rare and an increasing proportion of conjugations result in nonviable ex-conjugants; meanwhile, autogamy becomes more common both in separate individuals and in the form of selfing (cytogamy) during apparent conjugation. As senescence progresses, autogamy no longer produces viable clones, so that eventually after some months no further sexual processes of any type are successful and the ageing clone reaches a state of genetic death because its members can no longer participate in the foundation of a new clone. Ciliates in the senescent clone may still divide asexually and the clone may persist for some months in a state of reduced vigour before all of the population die and the clone suffers somatic death by ultimate loss of macronuclear function. It is important to recognize that both forms of sexual process, conjugation and autogamy, can result in rejuvenation and the foundation of a new clone; apparently the successful renewal of the macronucleus is the vital requirement in these processes. The clone of ciliates may be compared with a single multicellular animal in respect of ageing and ability to participate in sexual processes. Some ciliates do not show a comparable senescence, since in amicronucleate strains of *Tetrahymena* and *Didinium* conjugation is not possible, yet these strains have been maintained in culture for many years; it also appears that in one syngen of *P. aurelia*, which contains only one mating type and cannot normally conjugate, the same pattern of ageing cannot occur. The extrusion of DNA bodies from ciliate macronuclei has been widely reported, and may be concerned with the maintenance of macronuclear vitality.

Investigations of mating type and sometimes of clonal ageing have been made on a few other ciliates, including species of *Euplotes* and *Tetrahymena* as well as other species of *Paramecium*. Some differences are illustrated by *Paramecium bursaria*, where most of the six syngens that have been recognized contain more than 2 mating types. In one syngen there are eight mating types, and a mature member of one mating type will conjugate with a mature member of any of the other seven mating types of that syngen, but not normally with any member of any other syngen (see Table 4.2). In *P. bursaria* the immature period lasts several to many weeks and the period of maturity may last for many years, so that the complete life span of the clone may exceed ten years.

Table 4.2 The mating types of *Paramecium bursaria*. Data derived from T. Sonneborn.[244]

Syngen	Mating type	Will mate with types	Syngen	Mating type	Will mate with types
1	A	BCD	3	N	OPQ
	B	ACD		O	NPQ
	C	ABD		P	NOQ
	D	ABC		Q	NOP
2	E	FGHIKLM (R)	4	R	S
	F	EGHIKLM		S	R
	G	EFHIKLM	5	T	—
	H	EFGIKLM			
	I	EFGHKLM	6	U	VWX
	K	EFGHILM (R)		V	UWX
	L	EFGHIKM (R)		W	UVX
	M	EFGHIKL (R)		X	UVW

A complex pattern of **multiple mating types** has been described by Kimball for *Euplotes patella*.[135] In this species one syngen has six mating types, and mating has been found to be controlled by three substances secreted by the ciliates. A ciliate which produces substance 1 will conjugate in the presence of substances 2 or 3, a ciliate which produces substance 2 will conjugate in the presence of substances 1 or 3, and so on. Ciliates of some mating types produce two of the substances and mate only in the presence of the third. This system normally results in mating between ciliates of different mating types, but experimentally it is found that individuals of the same mating type will conjugate in the presence of an appropriate substance produced by members of another mating type. It is concluded that the pattern of mating illustrated in Table 4.3 is controlled by a set of three multiple alleles.

Control of the mating type resides in the macronucleus; the mating type of a clone is established at the formation of the new macronucleus following conjugation and is transmitted at macronuclear fission. However, the mating type of a new macronucleus is not necessarily determined by

Table 4.3 Genotypes of the mating types of *Euplotes patella*. Data derived from tables by R. Kimball.[135]

Mating type	Will mate in filtrate of mating types	Substances produced	Genotype
I	II, III, V	1, 2	mt^1, mt^2
II	I, V, VI	1, 3	mt^1, mt^3
III	I, II, IV, V, VI	3	mt^3, mt^3
IV	I, II, III, V, VI	1	mt^1, mt^1
V	I, II, IV	2, 3	mt^2, mt^3
VI	I, II, III, IV, V	2	mt^2, mt^2

genes derived from the micronuclear syncaryon, since the character of the cytoplasm in which the macronucleus develops can influence mating type. For example, a newly formed macronucleus of *P. aurelia* has the potentiality of producing either mating type.[244] Following conjugation the two ex-conjugants contain the same chromosome complement since

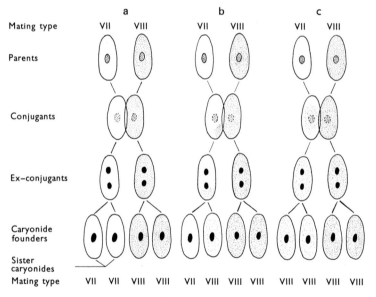

Fig. 4.11 Inheritance of mating type in Group B syngens of *Paramecium aurelia* exemplified by conjugation between individuals of mating types VII and VIII. In (**a**) the mating type of the two sister caryonides formed from each ex-conjugant is the same and corresponds to that of the parent conjugant, but in (**b**) and (**c**) following a lesser or greater transfer of cytoplasm from the type VIII conjugant to the type VII conjugant, three or all of the four caryonides formed are of the same mating type. Information from G. H. Beale.[19]

each syncaryon received identical haploid sets of chromosomes from the two parental cells. The macronuclei which develop in the two cells will therefore be formed from identical nuclei but will be surrounded by different cytoplasm. In one group of syngens (Group A) mating type seems to be determined randomly during the establishment of each macronucleus, so that the two macronuclei within the same ex-conjugant cell may determine different mating types in the cells produced at the first fission after conjugation. Within the clone derived from one ex-conjugant there may be two subclones (caryonides—clones with identical macronuclei) of different mating type, one for each of the newly constituted macronuclei; there is also random determination of mating type in the two caryonides established following autogamy. In the other group of syngens (Group B) cytoplasmic factors determine which nuclear characters shall be expressed, so that the mating type corresponds to that of the parental cytoplasm and the mating type of both sister caryonides within a clone will normally be the same (Fig. 4.11a). In exceptional cases of prolonged conjugation a large amount of cytoplasm may flow from one ciliate to the other through the conjugation pores, and the macronucleus which develops in the mixed cytoplasm may have the mating type of the parental cell or of the foreign cytoplasm (Fig. 4.11b, c).

Genetic significance of protozoan life cycles

The genetic mechanism of all organisms has built into it the possibilities of conservation and variation of information content. Conservation of information is related with the mechanism of accurate replication of the DNA code for genetic continuity, so that offspring are basically like their parents; prior to both mitosis and meiosis new genes are produced that are exact replicas of those already present. Variation is vital because it provides the raw material for natural selection to work on—the variants of pattern that can be 'tested' against the environment and may lead to an evolutionary change in the population if the variant has an advantageous feature.

The phenomenon of genetic variation has a number of facets that are less widely appreciated than the conservation of genetic information. The source of genetic change is mutation—an alteration in the molecular configuration of the gene which results in a modification of the gene product. Such a change may affect any one nucleotide or any combination of nucleotides within the DNA chain of the gene, so that a vast number of different mutations could affect any gene. A mutated gene may depend for its expression on the other genes in the cell, e.g. it may be a recessive gene masked by a dominant allelic gene, or it may form part of a complex of genes involved in producing enzymes for a chain reaction, and may only find expression if the other genes of the complex are present. In a haploid organism the first of these two could not occur, and the second would be more restricted since only one gene of each type would be present. It is therefore more likely that a gene in a haploid organism will find immediate expression for selection or rejection and so is less likely to remain hidden than in a diploid organism. Mutant genes that are not immediately beneficial may be more likely to survive in forms where the diploid phase predominates, so that genetic variability and the potentiality for genetic change are greater in populations of diploid Protozoa than in haploid Protozoa. On the other hand it is clearly advantageous to employ haploid forms such as *Chlamydomonas* in experimental studies of certain aspects of gene functioning, since mutant genes are likely to find immediate expression. The occurrence of a mutation in a favourable combination for expression will occur very rarely in an asexually reproducing organism, especially if it is haploid, so that changes will normally take place slowly unless very large populations of the organisms occur. From the point of view of the conservation of genetic variability, it is particularly advantageous for an asexual organism to be polyploid.

If the organism reproduces sexually the chances of the mutated gene finding itself in a favourable combination are vastly increased by the processes of meiosis and fertilization. During meiosis, not only do the chromosomes segregate independently, so that the different gametic nuclei produced by an organism contain a full range of different combinations of chromosomes from the set of pairs of homologous chromosomes of the parent, but the occurrence of new combinations of genes within chromosomes is made possible by the exchange of sections of chromosomes during crossing over. Every gametic nucleus may therefore contain a unique set of genes, and at fertilization two unique sets of genes come together, to make more likely the occurrence of a favourable combination of other genes in the same nucleus as a mutated gene. The spread of genetic variation through a population is greatly enhanced by the occurrence of sexual processes involving assortment of genes at a complete meiosis and recombination of genes from different parents at fertilization. The more remote the relationship between the parents, the greater the difference between their genetic constitutions, and therefore devices that encourage cross fertilization, such as the 'sexes' of chlamydomonad flagellates or the mating types of ciliates, will increase the range of new genetic combinations. Similarly the period of immaturity of the clones of *Paramecium* encourages out-breeding, and the longer the periods of immaturity and maturity the further the clone may spread before mating and the greater the chance that the ciliate will mate with an individual of an unrelated clone.

Reduction and loss of sexual processes

The genetic significance of sexual processes is reduced or lost in many Protozoa, as indicated in Fig. 4.12. The absence of crossing over in one-

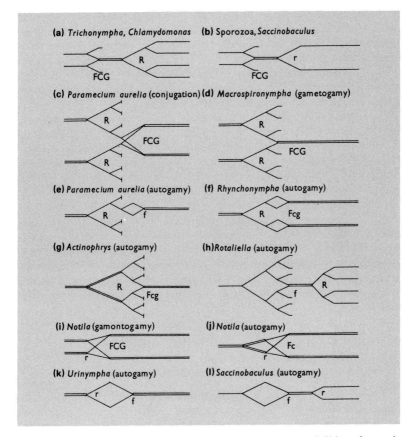

Fig. 4.12 Life cycles of some Protozoa to indicate the possibilities of genetic variation provided by sexual processes. R indicates a two-division meiosis involving crossing over as well as independent assortment of chromosomes; r indicates a one-division meiosis with assortment of chromosomes but no crossing over; FCG indicates fertilization that brings together new combinations of chromosomes and genes from different parents; Fcg indicates fertilization that brings together new combinations of chromosomes and genes derived from the same parent; Fc indicates fertilization that brings together new combinations of chromosomes from the same parent, but because of the one-division meiosis the gene content of the chromosomes remains unchanged; f indicates fertilization in which the genetic content of the two sets of chromosomes that come together is identical. Single lines indicate haploid stages and double lines diploid stages.

division meiosis of Sporozoa and some flagellates means a loss of the possibilities of gene reassortment and the production of new patterns of gene
linkage. The process of autogamy, as seen in *Paramecium aurelia*, retains
the advantages of meiosis in providing haploid nuclei with reassorted genes
but loses the advantage of fertilization because it brings together two
identical sets of genes, so that nuclei produced by autogamy of this type
are completely homozygous. In the heliozoans the situation is somewhat
different since the two pronuclei which fuse in autogamy are products of
different meioses, and homozygosity is unlikely to be complete in spite
of repeated autogamies; because of its special features this form of autogamy
is sometimes referred to as paedogamy.

The reduction in genetic significance of sexual processes in the complex
flagellates studied by Cleveland is particularly striking. The full extent
of genetic variation may occur in the haploid *Trichonympha* or diploid
Macrospironympha, but the variation is reduced because of one-division
meiosis in the gametogamy of *Saccinobaculus* or gamontogamy of *Notila*,
and it is reduced because of autogamy in *Rhynchonympha*. Although both
autogamy and one-division meiosis may occur together in *Notila*, the
process retains some genetic value because the gametic nuclei come from
different meioses in which independent segregation of chromosomes may
occur. Autogamy of *Urinympha* has no genetic significance since the
chromosomes which separate at the one-division meiosis are reunited at
fertilization, and similarly autogamous sexual reproduction of *Saccinobaculus* results in no genetic change since the zygote must be completely
homozygous.

DIVISION OF THE CELL AND MORPHOGENESIS

Timing of events in the life cycle of *Tetrahymena*

Division of the nucleus is normally followed by division of the cell.
These two events are frequently the most obvious stages of a process
which involves many parts of the cell and which commences before any
nuclear changes are seen. The most intensively studied protozoan life
cycle is that of *Tetrahymena pyriformis*, and the timing of some of the
events in this life cycle are shown in Fig. 4.13. This organism has been
found particularly suitable for these studies because large numbers of
ciliates may be obtained at the same stage in the life cycle by the appli-
cation of temperature shocks to synchronize cell division.[293]

Two transition stages appear to control the timing in this life cycle.
There is evidence that once a cell of *Tetrahymena* has successfully entered
the macronuclear S phase, preparations are set in motion for the next
division of the cell[205] (p. 48). These preparations involve not only the

synthesis of new DNA, but also the production of a variety of proteins, some of which are used in the replication of cell organelles prior to or subsequent to division. The first visible event is the duplication of ciliary basal bodies for the formation of new mouth ciliature. The original mouth of this ciliate is retained by the anterior daughter (proter) and a new mouth for the posterior daughter (opisthe) is formed just posterior to the equatorial fission furrow line and directly behind the existing buccal cavity (Fig. 4.14). These basal bodies first appear as an area of irregularly arranged kinetosomes which later become organized to form the bases of the three membranelles and the undulating membrane,[78] completion

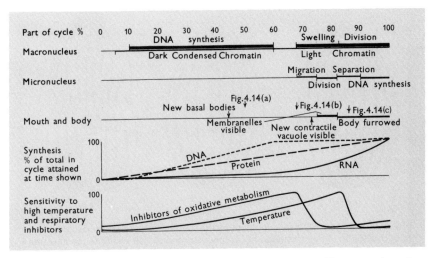

Fig. 4.13 Some events during the life cycle of *Tetrahymena*. The separation of two daughter cells occurs at time O, and the appearance of the cell at three stages during the cycle showing the development of the new mouth structures may be seen in Fig. 4.14. Information mainly from G. G. Holz.[114]

of these structures coinciding with the micronuclear mitosis. Late in the formation of the irregular field of kinetosomes the micronucleus migrates peripherally, the macronucleus swells and the ciliate becomes markedly less sensitive to inhibitors of aerobic respiration. A new contractile vacuole develops in the proter and begins to function during the period of micronuclear migration. Micronuclear mitosis is followed almost immediately by synthesis of new micronuclear DNA and by migration of the two micronuclei towards proter and opisthe; at this time macronuclear fission commences, and the appearance of an equatorial fission furrow quickly follows. The furrow is associated with a band of microfibrillar material, whose presence has been reported in many animal cells including several ciliates[129, 265] (Fig. 4.15). This microfibril band lies immediately below the pellicle and contracts to constrict the cell until only a narrow neck plugged by fibrillar material remains between the two daughters; the neck finally breaks and the two cells separate.

The second transition point may be demonstrated experimentally by heat treatment (34°C) or by such inhibitors of protein synthesis as puromycin and p-fluorophenylalanine.[292] Development of the oral ciliature is stopped or reversed by these treatments up to a critical time, at about the completion of the membranelles and before the undulating membrane

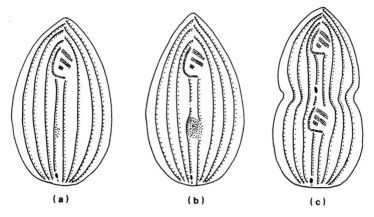

(a)   (b)   (c)

Fig. 4.14 Stages in the development of the mouth of the posterior daughter cell prior to division of *Tetrahymena*. In (a) the basal bodies that will form the new oral ciliature are beginning to multiply, in (b) they have formed an extensive 'anarchic field' and in (c) the basal bodies have been grouped to form the three membranelles and the undulating membrane. By the third stage the body is markedly furrowed and the new contractile vacuole pore of the anterior daughter may be seen.

is completed, after which these agents have no effect and the cell is fully committed to the division process. The evidence is consistent with the suggestion of E. Zeuthen that a special heat sensitive 'division protein' synthesized in these cells is required for morphogenetic activities connected with the preparations for division, and that at the second transition point this protein becomes insensitive to heat or is no longer required.

Control of the preparations for division is in fact the basis of the temperature treatment for synchronizing cell division of *Tetrahymena* reviewed by Zeuthen and Rasmussen.[292] Alternating periods of 30 minutes at 34°C and 30 minutes at 29°C permit cell growth, but hold the cells in an early stage of the preparation for division—the heat treatment prevents

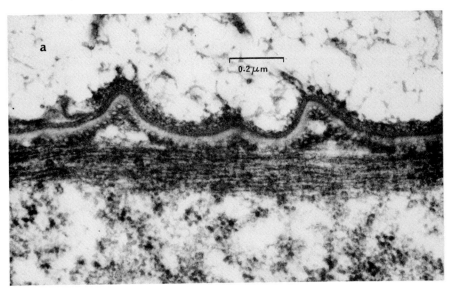

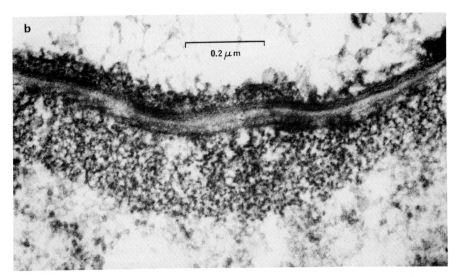

Fig. 4.15 Bundles of microfilaments beneath the cleavage furrow of *Nassula*, seen in longitudinal section (a) and transverse section (b). Micrographs by J. B. Tucker.[265]

differentiation and the period at optimum temperature (29°C) is not long enough for morphogenesis to progress as far as the onset of insensitivity to heat; at the end of a number of alternations of this type all of the cells are found to have an irregular field of kinetosomes and the micronuclear division is arrested in anaphase. If the animals are left at 29°C they will then pass through several cycles of division more quickly than usual (Fig. 4.16).

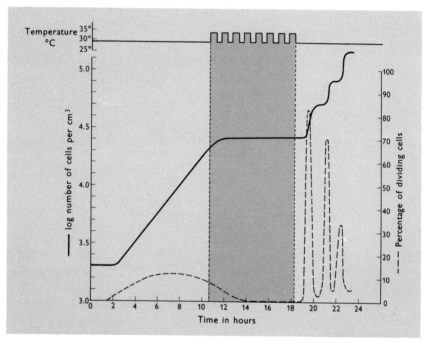

Fig. 4.16 The effect of repeated heat shocks on the population growth and percentage of dividing cells in a culture of *Tetrahymena*. From data by E. Zeuthen and O. Scherbaum.[293]

Zeuthen identified two cycles of activity in the life cycle of *Tetrahymena*; a cycle involving alternate periods of DNA synthesis and no synthesis of DNA, and a second cycle in which morphogenesis and cell division alternate. It would appear that under normal conditions these two cycles are interlinked, events in one cycle serving as signals for events in the other; perhaps the production of a specific protein follows macronuclear DNA synthesis and promotes morphogenesis, while certain stages in morphogenesis exert a blocking effect on macronuclear DNA synthesis.

The synchronizing temperature treatment prevents morphogenesis without preventing the DNA synthesis cycle, so that after the treatment the ciliates have an abnormally large store of DNA and may run through several divisions at a higher rate than usual before shortage of DNA and the need for time for DNA synthesis brings about a normal rephasing of the two cycles. The kinetosomes appear to play a particularly important morphogenetic role, and the state of development of the oral apparatus may be a key factor in the morphogenesis–cell division cycle; if the development of the oral primordium is blocked by a temperature shock, the existing structures are resorbed and replaced by the development of a new field of basal bodies and their differentiation to form the oral ciliature.

Morphogenesis in *Stentor*

A complete regeneration of mouthparts occurs in many ciliates if the fully formed oral ciliature is damaged or removed. This has been studied in detail by V. Tartar in *Stentor coeruleus*,[254, 255, 257] where the development of a new mouth primordium occurs in a specific region of the lateral surface at a place where closely spaced kineties meet widely spaced kineties (the region of pigment-stripe contrast); grafting experiments showed both the specificity of this site and the nature of the stimulus which leads to primordium formation. It appears from this work and that of N. de Terra[56] that an oral primordium will develop when the ratio of oral structures to body size falls below a certain level; the primordium development may lead to body reorganization, regeneration or division, and the newly formed membranelles may be added to the existing membranelles of a small row to extend it, may form a complete new mouth ciliature for a damaged individual or may form the feeding ciliature for the posterior daughter in division. The determination of whether primordium formation in an undamaged cell results in reorganization or division is a function of cell size—only larger cells divide. Following the appearance of a primordium, the macronucleus always undergoes reorganization by coalescence of nodes and subsequent renodulation; these processes are shown by nuclear transplantation experiments to be under cytoplasmic control. If the primordium formation leads to constriction of the cell, the macronucleus is divided by the fission furrow while it is in the condensed state, but if the nucleus is not constricted, either because the cell is regenerating rather than dividing, or because the macronucleus has been artificially displaced to one side of the fission line, it does not divide. The degree of control over nuclear events exerted by the cytoplasm is striking, and the ultimate source of the regulation appears to depend upon the spatial relationships of organized structures in the pellicle. It was found in a hypotrich ciliate that the removal of one cirrus led to complete reorganization of the body ciliature. Many aspects of morphogenesis in Protozoa have been reviewed by Tartar.[256]

Specialized and modified cell cycles

Although the cell cycle has not been studied in comparable detail in other Protozoa, it is believed that similar controls and interactions operate in cells of other types, and that these may be modified in a variety of ways in more complex cell cycles, e.g. where these involve encystment (p. 260). Evidence of preparation for division before nuclear changes are seen is available for such amoebae as *Euglypha* where plates for the shell of the new daughter accumulate in the cytoplasm of the parent cell (Fig. 7.10f, p. 168), and also in such flagellates as *Euglena* where new pellicle ridges appear between existing ones before division of the nucleus. Where centrioles are present these divide before changes are seen in the nucleus, and in flagellates where the centrioles are normally closely connected with the flagellar apparatus the flagellar basal bodies and associated organelles are usually duplicated before nuclear division starts.

The replacement or renewal of cytoplasmic structures at binary fission is a common event in both flagellates and ciliates. In such flagellates as *Macrotrichomonas* and *Lophomonas* new sets of body organelles are formed in both daughters at binary fission, and the organelles of the parent are resorbed. In the ciliate *Nassula* the original feeding structures are broken down and a new cytopharyngeal complex is formed in each daughter.[264] At binary fission of hypotrich ciliates new locomotory cirri and feeding organelles develop in both daughter cells.[268] In these cases the morphogenetic cycle somehow controls both the breakdown and the synthesis of the same type of structure simultaneously.

The control system of the cell cycle must also be modified in those cases where the protozoon undergoes multiple fission. A class of division that is sometimes referred to as syntomy occurs when the nucleus divides repeatedly and then a body of cytoplasm gathers around each nucleus to form a daughter cell. Such reproduction is common in Sporozoa, e.g. in the formation of gametes of *Monocystis*, in the formation of sporozoites in *Plasmodium*, and in the formation of merozoites at schizogony of *Plasmodium*. A comparable schizogony occurs in *Trypanosoma lewisi* in the gut cells of the rat flea, and in the multiple division of foraminiferans and such parasitic amoebae as *Entamoeba*. This form of multiple division is distinguished from the rapid sequence of binary divisions (called palintomy) shown by such ciliates as *Ichthyophthirius*, where a giant ciliate encysts and undergoes many fissions without growth to produce numerous small ciliates. In the life cycle of this parasitic ciliate periods of growth during which the reproductive cycle is inhibited alternate with periods of repeated reproduction when no growth takes place. Palintomy is also found in flagellates, where the formation of gametes, e.g. the microgametes of *Volvox*, may involve this pattern of division.

Budding is a form of fission in which the parent cell may be only slightly modified in the process of reproduction. In simple budding of a suctorian, e.g. *Paracineta*, or a chonotrich, e.g. *Spirochona*, the sedentary parent ciliate retains its feeding apparatus during nuclear division and the production of a terminal or lateral bud which develops into a ciliated larva. Multiple budding of 4 to 12 larvae simultaneously is found on the suctorian, *Ephelota*, while in other suctoria, e.g. *Acineta*, the larva may be budded off within a brood pouch formed by an invagination of the body surface. The formation of linear chains of ciliates by budding is found in the order Astomatida.

5

The Flagellate Protozoa

Unicellular flagellate organisms are frequently encountered in fresh and salt water and as symbionts. In addition to pigmented and colourless trophic cells, such flagellate unicells may be algal zoospores or gametes, or possibly the spermatozoa of higher animals. While the last are likely to be identifiable as gametes because of their reduced structure, the zoospores and gametes of algae are often very similar to flagellate trophic cells in the same group or a closely related group. Within many algal groups there are often coccoid, amoeboid, colonial, filamentous and sometimes thalloid organisms as well as the independent unicellular flagellates.[216] Information on the distribution of these life forms is given in Table 5.1, but details of the structure of multicellular algae and of the life cycles of multicellular algae which possess flagellate stages will not be mentioned further in this book, where attention will be confined to the flagellate individuals.

It is implicit in the comments on pp. 6–9 that the primary evolutionary divergence within the Contophora did not involve a separation of animals from plants, but took place within the autotrophic forms before the establishment of heterotrophic eucaryotes. Heterotrophic forms arose later in most of the lines of autotrophic organisms, giving rise to unpigmented flagellates, amoebae and other Protozoa, fungi and higher animals. In particular it should be stressed that several groups of heterotrophic flagellates are more closely related to autotrophic forms than they are to other groups of heterotrophic organisms. Therefore, an integrated classification of all flagellate groups is favoured, and should supersede the division of convenience into phytoflagellates and zooflagellates, which is clearly not a natural division.

Problems of nomenclature and rank are inevitable in the integration of classifications used by workers with different viewpoints. The algal classes of botanists are generally equivalent among the flagellates to the protozoan orders of zoologists, and botanical and zoological codes of nomenclature use different endings for the higher taxa. It is not appropriate in this book to propose a new nomenclature for flagellate classification but merely to give an account of the range of form and suspected relationships of flagellates. Some integration should be achieved by considering the inter-relations of flagellates and by giving both botanical and zoological names for comparison (Table 5.1), although in the text only one of these names will normally be used. Some of the taxa of flagellates have only recently been described in specialist journals.

All groups of the Contophora except Phaeophyceae and Bacillariophyceae have vegetative flagellate cells; flagellate gametes occur in both of these groups and flagellate zoospores also occur in Phaeophyceae. In all algae except the Rhodophyceae, therefore, it is possible to study the structure of flagellate cells and to compare them with each other and with the heterotrophic flagellates. New information is now available about the chemistry of pigmentation and food storage products and about the ultrastructure of such cell components as chloroplasts and flagella, all of which are regarded as valuable features in classification. This information is the main basis of Table 5.1, and some guide to the significance of the data will be given in the following paragraphs.

STRUCTURES IN FLAGELLATE CELLS

Chloroplasts and eyespots

A major subdivision of the Contophora separates a chromophyte series of forms without chlorophyll *b* from a chlorophyte series which possess chlorophyll *b*. Some of the chromophytes possess chlorophyll *c*. Few groups contain α-carotene, while all pigmented groups except Cryptophyceae have β-carotene. The xanthophylls are often specific to a particular group of algae, but are reported to be difficult to separate and identify. The light energy absorbed by these carotenes and xanthophylls is thought to be passed to chlorophyll for use in the photosynthetic process.

The fine structure of chloroplasts shows an interesting range of variation (Fig. 5.1). Chloroplasts consist basically of thylakoids embedded in a matrix and enclosed by a double unit membrane (p. 20). In Rhodophyceae many single separate thylakoids occur within the double chloroplast membrane (type a). Thylakoids adhere together in pairs to form the lamellae in the chloroplasts of Cryptophyceae (Fig. 5.2), and the double chloroplast membrane is enclosed by two endoplasmic reticulum membranes, the outer of which is continuous with the outer nuclear membrane, so that

Table 5.1   Features of

Group	Pigmentation				Plastid membranes				Stigma		Flagella	
	Colour	Colourless members	Chlorophylls	Carotenes	Thylakoids/lamella (usual number)	Girdle lamella present	Number of membranes enclosing plastid	Chloroplast e.r. continues around nucleus	Eyespot in plastid	Eyespot associated with flagellum	Number of flagella	Flagella similar
Rhodophyceae	R	×		α β	1	√	2	−	−	−	0	−
Cryptophyceae	Gr R Br Bl	√		α	2	×	4	√	√	×	2	Nearly
Haptophyceae	Y-Br	√	a		3	×	4	√			2 +H	√
Choanoflagellida	−	√	−	−	−	−	−	−	−	−	1	−
Bicoecida	−	√	−	−	−	−	−	−	−	−	2	×
Chrysophyceae	Y-Br	√	a	β	3	√	4	√	√	√	2	×
Ebriida	−	√	−	−	−	−	−	−	−	−	2	×
Xanthophyceae	Y-Gr		a	β	3	√ ×	4	√	√	√	2	×
Eustigmatophyceae	Gr		a	β	3	×	4	×	×	√	1(2)	−
Rhaphidiophyceae	Gr		a	β	3	√	4	×			2	×
Phaeophyceae	Br	×	a, c	β	3	√	4	√	√	√	2	×
Silicoflagellida	Y-Br		a, c		3	√	4	?			1	−
Bacillariophyceae	Y-Br	×	a, c	β	3	√	4	√	−	−	(1)	−
Dinophyceae	Br	√	a, c	β	3	×	3	×	×	×	2	×
Kinetoplastida	−	√	−	−	−	−	−	−	−	−	2(1)	×
Euglenophyceae	Gr	√	a, b	β	(2–6) 3	×	3	×	×	√	2(1, 3)	×
'Loxophyceae'	Gr		a, b					√	√	×	1(4)	−
Prasinophyceae	Gr		a, b		2–4	×	2	√	√	×	2, 4	√
Chlorophyceae	Gr	√	a, b	β	2–6	×	2	√	√	×	2, 4, many	√
Metamonadida	−	√	−	−	−	−	−	−	−	−	4 to thousands	√

Key to abbreviations: R, red; Gr, green; Br, brown; Bl, blue; Y, yellow; H, haptonema; C, coccoid;
form; Gam, gamete; √, present; ×, absent; −, character not expected and space, character unknown

various flagellate groups

One flagellum with 2 rows stiff hairs	Intraflagellar rod	α 1–4 glucan or β 1–3 glucan	Fat	Cell wall or skeleton material	Trichocysts	Unicellular forms	Multicellular forms	State(s) of flagellate form(s)	Example	Alternative name of group
—	—	α	√	'Agar'		C	Fil Th	—		
√ ?	×	α	√	—	√	C Fl A C	—	Veg	*Cryptomonas*	Cryptomonadida
×	×	β	√	Scales		Fl	Fil		*Chrysochromulina*	Coccolithophorida (in part)
×	×		√	Silica lorica		Fl	—	Troph	*Codonosiga*	Craspedomonadales
√ ?				? lorica		Fl	—	Troph	*Bicoeca*	Bicosoecida
√	×	β	√	lorica of silica cellulose		A C Fl	Fil	Veg Zoosp	*Ochromonas*	Chrysomonadida
			√	Intracellular silica		Fl	—	Veg	*Ebria*	
√	×		√			A C Fl	Fil Th	Veg Zoosp	*Bumilleria*	Heterochlorida
√	×		√			Fl	—	Veg	*Vischeria*	
√	×		√		√	Fl	—	Veg	*Vacuolaria*	Chloromonadida
√	×	β	√	Alginic acid			Fil Th	Zoosp Gam		
		β ?		Extracellular silica		Fl	—	Veg	*Dictyocha*	
√	×	β	√	Silica cell wall		C	Fil	Gam		
×	√	α	√	Internal cellulose plates	√	A C Fl	Fil	Veg Zoosp Gam	*Peridinium*	Dinoflagellida
×	√			Reinforced pellicle		Fl	—	Troph	*Bodo*	
×	√	β		Reinforced pellicle	√	Fl	—	Veg	*Euglena*	Euglenida
×	×	α				Fl	—	Veg	*Pedinomonas*	
×	×	α		Scales		Fl	—	Veg	*Pyramimonas*	
×	×	α	√	Cellulose cell wall		C Fl	Fil Th	Veg Zoosp Gam	*Chlamydomonas*	Volvocida
×	×	α		—		Fl	—	Troph	*Trichonympha*	

Fl, flagellate; A, amoeboid; Fil, filamentous; Th, thalloid; Veg, vegetative; Zoosp, zoospore; Troph, trophic

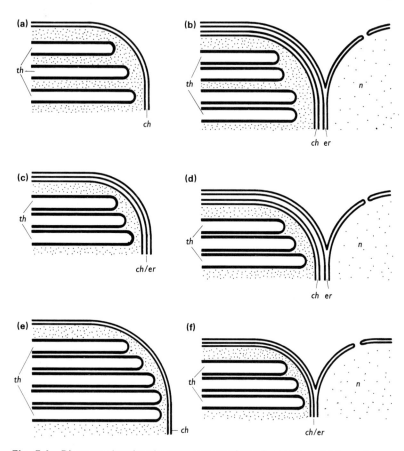

Fig. 5.1 Diagrams showing the arrangement of membranes in a variety of chloro-plasts described in the text; a, in Rhodophyceae: b, in Cryptophyceae: c, in Euglenophyceae: d, in Chrysophyceae: e, in Chlorophyceae: f, in Dinophyceae. *ch* Are membranes bounding the chloroplast, *th* are thylakoid membranes, and *er* are endoplasmic reticulum membranes which enclose the complete chloroplast and sometimes continue around the nucleus *n*.

the chloroplast is surrounded by four membranes (type b). All other chromophytes normally have three thylakoids in each lamella of the chloro-plast (e.g. Fig. 5.3), although in Dinophyceae and sometimes in Phaeo-phyceae four or more thylakoids may be present in each lamella. Most of these groups also have four chloroplast membranes, usually with the outer-most one continuous around the nucleus (type d); the reduction to three membranes seen in Dinophyceae may be the result of fusion of the two middle membranes (type f).[59] In the chlorophytes the number of thylakoids in each lamella is variable and usually larger than three (type e); the development of grana—discs within the chloroplast formed from localized stacks of thylakoids—is found in Prasinophyceae and Chlorophyceae, and two membranes enclose the chloroplast in all chlorophytes except Euglenophyceae where there are three chloroplast membranes (type c). A further

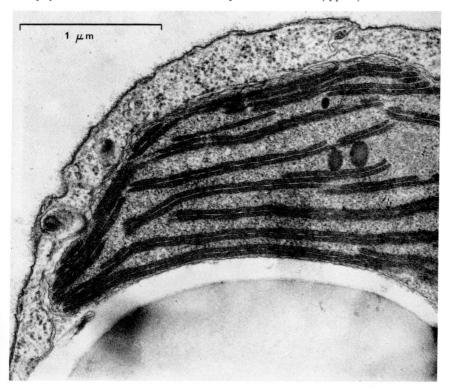

Fig. 5.2 Electron micrograph of a section through part of the chloroplast of *Chroomonas mesostigmatica* (Cryptophyceae) showing the paired thylakoids characteristic of this group. Micrograph by J. D. Dodge.[60]

ultrastructural feature of the chloroplast that appears to be of taxonomic value is the presence of a girdle lamella enclosing the other chloroplast lamellae. Relatively few members of some groups have yet been studied in sufficient detail for predictions to be made of the ultimate taxonomic importance of these chloroplast characters, but the groupings characterized by these features are consistent with divisions based on other characters.[103]

Pyrenoids are areas of dense matrix found within chloroplasts and associated with the formation of polysaccharides; they are often surrounded or capped by large grains of starch, or some other polysaccharide. Normally the pyrenoid occurs within the body of the chloroplast, but may be

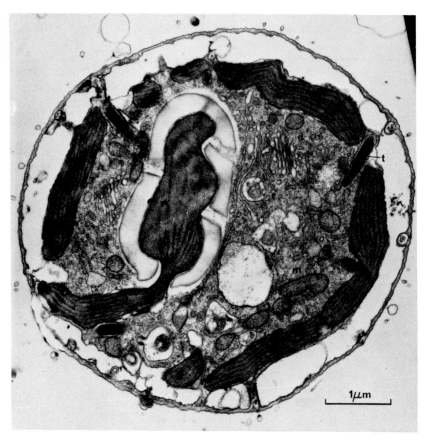

Fig. 5.3 Electron micrograph of a section through the dinoflagellate *Amphidinium carteri*, showing the chloroplasts (c) with lamellae containing three thylakoids and the pyrenoid (p) penetrated by lamellae and capped by a starch deposit. Mitochondria (m), dictyosomes (d) and trichocysts (t) are also present. Micrograph by J. D. Dodge and R. M. Crawford.[62]

a stalked projection from it (e.g. Fig. 5.3); in some cases the pyrenoid may be penetrated by tongues of cytoplasm or even of nuclear material. The pyrenoid is normally penetrated by membranous lamellae of the chloroplast; such lamellae may be of normal constitution or may have reduced numbers of thylakoids. The functional implications of variations of these features and the consistency of their appearance in the various groups are not known.[103]

Clusters of lipid globules containing orange or red carotenoid pigments are referred to as eyespots or stigma structures; they are common in pigmented flagellates and may show several patterns of relationship with chloroplasts and flagella (Table 5.1).[103] In most groups where it is present

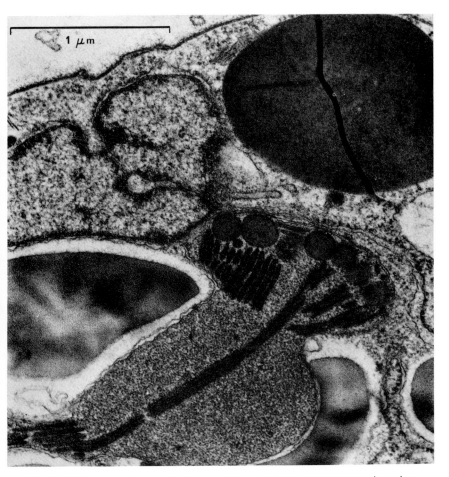

Fig. 5.4 Electron micrograph of a section of *Chroomonas mesostigmatica* (Cryptophyceae) showing part of the chloroplast containing the simple stigma of this flagellate; the pyrenoid region is enclosed by a starch deposit lying between the inner and outer pairs of chloroplast membranes. Micrograph by J. D. Dodge.

the eyespot is found within a chloroplast (Fig. 5.4), and may be associated with a short or backwardly directed flagellum or may be distant from the flagella. Where the eyespot is situated outside the chloroplast, it may still be associated with the basal region of a flagellum, in this case an active

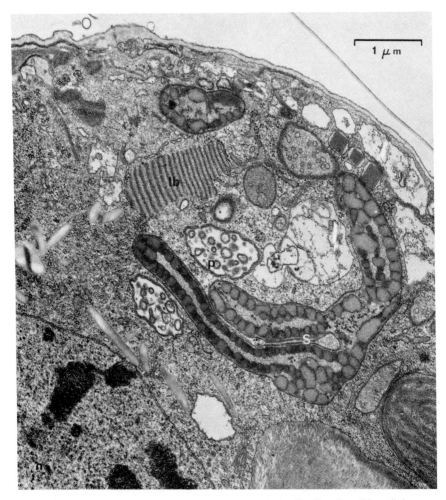

Fig. 5.5 Electron micrograph of a section through the dinoflagellate *Glenodinium foliaceum* showing part of the complex stigma (s) and a lamellar body (lb), both of which are closely associated with the flagellar bases (not shown here). Parts of the pusule system (p) and nucleus (n) also appear in the section. Micrograph by J. D. Dodge and R. M. Crawford.[63]

locomotory one. The most complex eyespot structures are found in some Dinophyceae (e.g. *Pouchetia*) where the stigma is not associated with either chloroplasts or flagella, and may be a large structure with several components, while in some dinoflagellates, e.g. *Glenodinium* (Fig. 5.5), the stigma is simpler and lies closer to the flagellar bases.[63] The eyespot is assumed to be concerned with orientation to light and phototactic behaviour, but the reason for the extreme development of the structure in some Dinophyceae is obscure.

Flagella and locomotion

The ultrastructure of flagella and the way in which they propel the body are still imperfectly known for some groups, although if one or other of these is known it is often possible to infer the nature of the other. Flagellar ultrastructure varies in the form and disposition of hairs and scales on the flagella,[169] and in the presence in some groups of an intra-flagellar rod lying alongside the axoneme; these variants are illustrated in Fig. 5.6, and their occurrence in various groups is indicated in Table 5.1.

Very commonly two flagella are present, and where only one emergent flagellum is visible, derivation from a biflagellate ancestor can usually be inferred by the presence of a second internal flagellum or by a close relationship with a biflagellate group. A particular exception to this is the Choanoflagellida, in which the flagellum acts as a pulsellum, pushing water away from the body. The two flagella of biflagellate forms may differ in both structure and activity or may be similar and used in a similar way in locomotion, as in many Haptophyceae, some Chlorophyceae and probably in many Cryptophyceae, where however there is some difference in hair pattern on the two flagella (Fig. 5.8). Heterodynamic flagella are of two main types. That exemplified by the Chrysophyceae, but also found in Bicoecida, Xanthophyceae, Eustigmatophyceae, Phaeophyceae, some fungal zoospores and possibly in Ebriida involves the use of an anteriorly directed flagellum with two rows of thick lateral hairs as a tractellum (drawing water towards the body), while the second, usually less active, flagellum may be short, or, if it is longer, it is directed back as a means of attachment or trailing flagellum. Since this second flagellum is often closely associated with the eyespot, it may determine the direction of locomotion. In the Silicoflagellida and some members of the Chryso-phyceae and Eustigmatophyceae only a single flagellum is present, and this appears to be homologous with the anterior hairy flagellum of typical members of the Chrysophyceae. The other groups showing heterodynamic flagella are more diverse, but share the possession of an intraflagellar rod and the occurrence of a unilateral row of hairs on at least one flagellum.[150] The use of the flagella is also diverse, for the Dinophyceae normally use

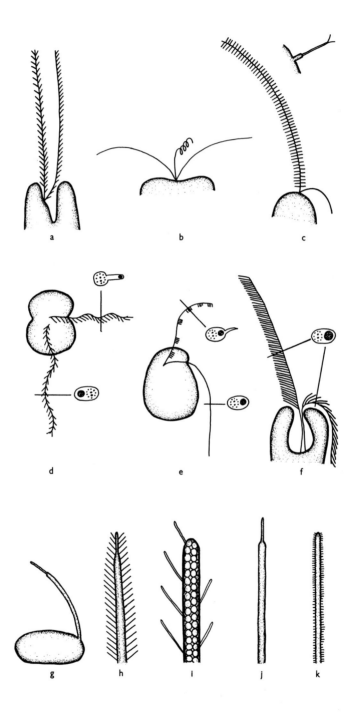

a   b   c

d   e   f

g   h   i   j   k

one longitudinal flagellum covered with short hairs as a pulsellum (p. 36), while a transverse flagellum with unilateral thin hairs moves in a groove in a unique manner, causing both rotation and forward movement (Fig. 2.12, p. 36); among the Kinetoplastida the bodonids use one anterior flagellum with unilateral tufts of short hairs as a tractellum while the second flagellum trails behind, and the trypanosomids mostly have a single smooth emergent flagellum which serves as tractellum or pulsellum (Fig. 2.12); finally the Euglenophyceae show a variety of patterns in which a flagellum with unilateral long hairs normally emerges from the anterior end of the body but is turned backwards to a greater or lesser extent in motion (Fig. 2.12). The algae tentatively included in the Loxophyceae all appear to have flagella with a thin tip, and may have two rows of very fine hairs, while the two or four flagella of the Prasinophyceae have scales and usually some characteristic thick hairs that are easily shed. The flagella of the Chlorophyceae are smooth or covered by a tomentum of soft hairs, and occasionally have delicate scales. Members of these last three groups of green flagellates have a characteristic stellate pattern of fibrous inter-connections between the 9 peripheral fibres of the flagellum at the level of the transition zone between basal body and flagellum (Fig. 2.9, p. 29); this pattern may be found in other flagellates and in cilia, but is less well developed elsewhere. One or two pairs of flagella are usually present on members of the Chlorophyceae, and numerous pairs of flagella occur on the zoospores of some green algae. The majority of groups of symbiotic colourless flagellates are also characterized by an increase in flagellar numbers to four or eight in many simple forms and up to many thousands in complex forms.

Internal structures attached to flagellar bases

Striated fibres or bundles of microtubular fibres frequently connect the bases of flagella with each other or pass from the flagellar bases to the cell surface or to other cell structures.[169] It is not clear from the evidence now available from members of various groups of pigmented flagellates whether the characteristics of root systems will be useful in taxonomy. In flagellates of both Chromophyta and Chlorophyta groups of 2–5 or more microtubular fibrils, often associated with a striated fibre,

Fig. 5.6 Diagrams showing the arrangement of flagella and hair patterns on flagella of examples from various flagellate groups; a, Cryptophyceae: b, Hapto-phyceae : c, Chrysophyceae (with inset showing the three regions of a single hair, c.f. Fig. 2.13): d, Dinophyceae: e, Kinetoplastida (*Bodo*): f, Euglenophyceae: g and h, Loxophyceae: i, Prasinophyceae: j and k, Chlorophyceae. The trans-verse sections in d, e and f show the presence of intraflagellar rods running along-side the axoneme in these forms.

have been found to run from the flagellar base to the cell surface and pass around much of the cell immediately under the cell membrane, where they appear to make contact with the chloroplasts, nucleus or other organelles. Other rootlets, often striated fibres, but occasionally bundles of microtubules, run down into the cell and usually make intimate contact with the surface of the nucleus. These root fibres are well situated to serve the function of anchorage. In the Euglenophyceae, Kinetoplastida and Dinophyceae microtubular fibrils pass from the flagellar bases to the cell surface where they appear to mingle with the microtubules which form an extensive pellicular array in most of these forms (Fig. 5.14, 5.22 and 5.26). Complex relationships between flagellar bases and the nuclei and other cell organelles are found in metamonad flagellates; in *Tricho-monas* for example there is no longer a pellicular system of microtubules, but these appear to have moved inwards to form an internal cylinder, the axostyle, which closely surrounds the nucleus and is associated anteriorly with the base of one of the flagella.[108] The bases of other flagella are associated with a contractile fibre called the costa and a second striated fibre that makes connections with numerous 'parabasal' golgi complexes (Fig. 5.30). In these flagellates the flagella and associated structures are referred to as the caryomastigont system, which may be duplicated, multiplied or modified in other metamonads. Clearly the structures attached to these flagella serve diverse functions in addition to anchorage. In simple and complex flagellates the basal bodies or structures attached to them commonly provide the terminations of the nuclear spindle in mitosis.

Extracellular materials and skeletal structures

Many free-living flagellates produce some extracellular materials which can be regarded as having a protective function. Most obvious amongst these are the cell walls of the Chlorophyceae, made of a cellulose-like polysaccharide. Scales occur on the outside of the cell membrane in the Prasinophyceae, Chrysophyceae and Haptophyceae, and in the last of these calcareous material may be deposited on the scales.[172] A lorica of secreted material is found in Bicoecida, in many Choanoflagellida, some Chrysophyceae and a few Euglenophyceae. The strengthening of the body surface may also be achieved by pellicular thickenings and addition of microtubules within the membrane, as in Dinophyceae, where cellulose plates are found within internal membranous alveoli that are underlain by microtubules (Fig. 5.14), and in Euglenophyceae, where thickened pellicular strips and microtubules are arranged longitudinally (Fig. 5.22). Internal silica skeletons are found in Ebriida, while the silica skeletons of Silicoflagellida are at least partly external and of different construction; endogenous siliceous cysts are formed by some Chrysophyceae. The cell walls and other extracellular material of multicellular forms are often more substantial, but are not considered here.

Food storage substances

Food is generally stored as carbohydrate granules or lipid droplets. Two forms of carbohydrate are widely distributed; α-1:4 glucans as starch and glycogen and β-1:3 glucans as leucosin, laminarin and paramylon. Starch is stored within the chloroplasts of Prasinophyceae and Chlorophyceae and between the chloroplast membranes and the surrounding endoplasmic reticulum envelope in Cryptophyceae (Fig. 5.4); in other groups the polysaccharide is normally found in the cytoplasm (e.g. Figs. 5.3, 5.15). Lipid droplets are found in the cells of most groups, and in some cases form the only food reserve.

FEATURES OF THE FLAGELLATE GROUPS[77, 80, 90]

Several members of the **Cryptophyceae**, including the pigmented *Cryptomonas* and the colourless *Chilomonas* (Fig. 5.7a), are exceedingly common freshwater organisms, and *Hemiselmis* (Fig. 5.7c) is a marine form. The group appears isolated from other chromophytes by the simplicity of the chloroplast and characters of the flagella, stigma and food reserves. The two flagella (Fig. 5.8) emerge from an anterio-lateral invagination, the walls of which contain numerous trichocysts and the opening of the contractile vacuole (Fig. 5.7a). Sometimes blue and red pigments are present in addition to the normal brown–green of the typical chloroplast pigments.

The **Haptophyceae** have only recently been recognized as a separate group on the basis of the possession of a haptonema (p. 32) and two smooth flagella; *Isochrysis* (Fig. 5.7h), with two smooth flagella but no visible haptonema, has also been placed in this group. They are clearly distinct from the Chrysophyceae with which they were formerly classified. As abundant members of the nannoplankton, they are important primary producers in the sea, and their abundance in the past can be seen by the enormous deposits of cretaceous chalk which are largely derived from the coccoliths formed by some of these flagellates. Coccoliths are calcified areas of the body scales[172] (e.g. Fig. 5.7e, f); they are diverse in shape and have been used as the basis for a classification of coccolithophore flagellates, but the finding that coccoliths formed on the motile stage of one flagellate (*Coccolithus pelagicus*) were different from those formed on the non-motile coccoid stage of the same flagellate throws doubt on this earlier work.[194] The scales of several species of *Chrysochromulina* (e.g. Fig. 5.9) studied by M. Parke and I. Manton[195] were found not to be calcified.

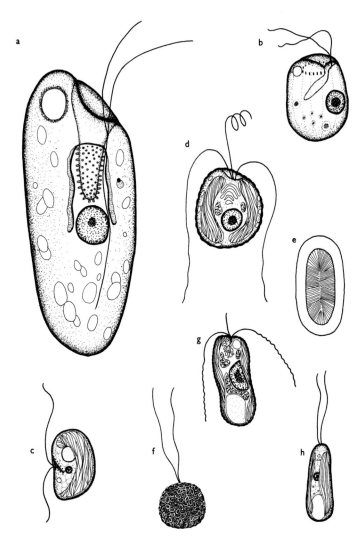

Fig. 5.7 Some representative members of the Cryptophyceae (a–c) and the Haptophyceae (d–h). a, *Chilomonas paramecium* (30 μm): b, *Cyathomonas truncata* (20 μm): c, *Hemiselmis rufescens* (6 μm): d, *Chrysochromulina kappa* (6 μm): e, scale of *Cricosphaera carterae* (length 2 μm): f, *Cricosphaera carterae* (15 μm): g, *Prymnesium parvum* (6 μm): h, *Isochrysis galbana* (5 μm).

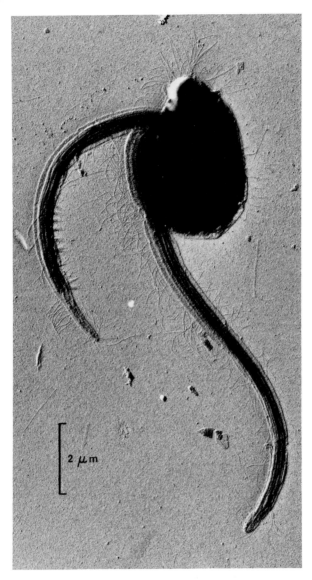

Fig. 5.8 Electron micrograph of a member of the Cryptophyceae (probably *Chroomonas* sp.) showing the implantation of the flagella and hair patterns on the flagella and body. Micrograph by J. D. Dodge and R. M. Crawford (unpublished).

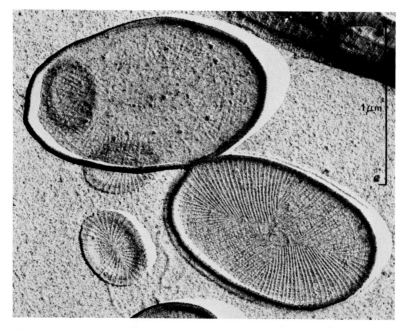

Fig. 5.9 Electron micrograph of scales of *Chrysochromulina* (Haptophyceae). Micrograph by R. M. Crawford (unpublished).

Members of this genus (Fig. 5.7d) are phagotrophic as well as autotrophic and occasionally an individual lacking plastids is found; these species have a non-flagellate phase which is amoeboid and also shows phagotrophy as well as autotrophy. *Prymnesium parvum* (Fig. 5.7g) is a minute member of this group which produces an extracellular toxin, supposedly proteinaceous, which is lethal to fish.

The **Chrysophyceae** form a very diverse group, which, according to many authorities, may have been the source of several other flagellate and

---

Fig. 5.10 Some representative members of the Chrysophyceae and a variety of small flagellate groups; a, *Oikomonas termo* (15 μm): b, *Chrysamoeba radians* (10 μm): c, *Synura uvella* (cells 30 μm): d, *Dinobryon sertularia* (cells 30 μm): e, *Anthophysa vegetans* (cells 5 μm): f, *Pedinella hexacostata* (10 μm): g, *Actinomonas mirabilis* (10 μm): h, *Codonosiga botrytis* (15 μm) (Choanoflagellida): i, *Salpingoeca fusiformis* (15 μm) (Choanoflagellida): j. *Bicoeca exilis* (12 μm) (Bicoecida): k, *Dictyocha speculum* (40 μm) (Silicoflagellida): l, *Hermesinum adriaticum* (Ebriida): m, *Bumilleria sicula* zoospore (15 μm) (Xanthophyceae): n, *Vischeria punctata* zoospore (10 μm) (Eustigmatophyceae): o, *Goniostomum semen* (50 μm) (Rhaphidiophyceae).

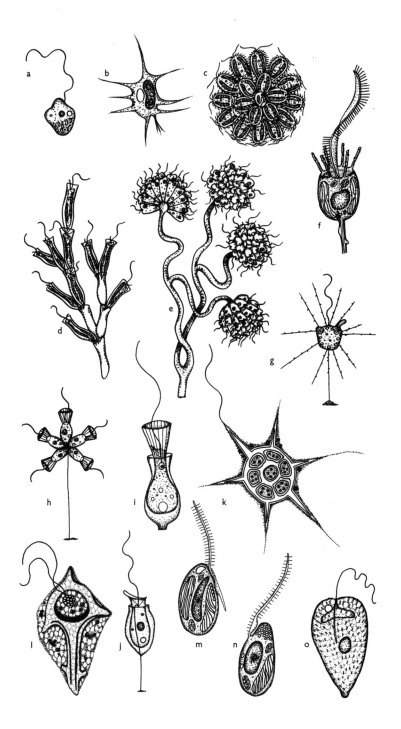

amoeboid groups. It includes a variety of pigmented and colourless, flagellate and amoeboid, solitary, colonial, plasmoidal and filamentous forms (Fig. 5.10a–g). The one or two chloroplasts usually have a golden-brown colour. The flagellate forms characteristically have two flagella. One is directed forwards and bears two rows of stiff lateral hairs (Fig. 2.13), and the second is smooth, is associated with the eyespot, is often shorter and may be directed backwards; the smooth flagellum is sometimes missing. They are often fresh-water forms, frequently with a sheath, lorica or scales. Many of the species combine autotrophic nutrition with phagotrophy by means of pseudopodia, sometimes by collecting particles

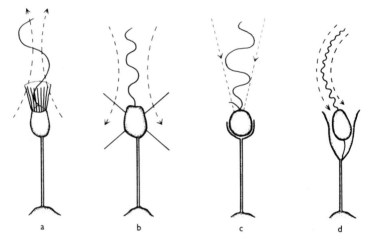

a   b   c   d

Fig. 5.11   Diagrams comparing the form of flagellar beating and the pattern of water flow around the flagella of (a) a collar flagellate, (b) *Actinomonas*, (c) a chrysomonad like *Ochromonas* and (d) *Bicoeca*. (Redrawn from Sleigh.[233]).

from water currents created by the anterior flagellum (Fig. 5.11c). Two possible evolutionary derivatives from such species are the **Bicoecida** (Figs. 5.10j, 5.11d) (loricate biflagellates in which one flagellum attaches the flagellate within the lorica while the second collects food), which are traditionally classified separately by zoologists, but should perhaps be regarded merely as colourless members of the Chrysophyceae, and the **helioflagellates** (p. 180), among which the colourless *Actinomonas* (Figs. 5.10g, 5.11b) shows similar features to the pigmented chrysophycean *Pedinella*[251] (Fig. 5.10f) in the possession of pseudopodia and a contractile stalk and may represent a link between the Chrysophyceae and the centrohelid heliozoans (p. 177).

    The relationships of the **Choanoflagellida** are less clear; although the

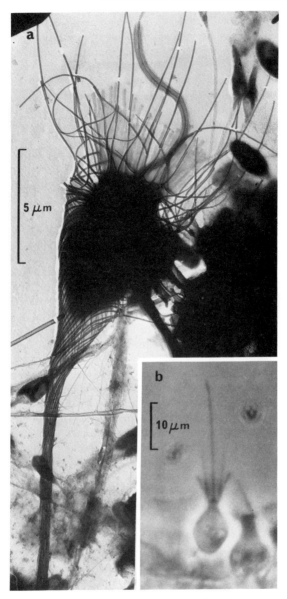

Fig. 5.12   a, Electron micrograph of the choanoflagellate *Acanthoeca spectabilis* from marine plankton. The dense cell body at the centre bears the flagellum and the collar filaments at its top and is enclosed by a lorica formed from siliceous rods. Micrograph by B. S. C. Leadbeater.[148] b, Photomicrograph of *Salpingoeca* sp. from fresh water.

collar of fine pseudopodia in these collar flagellates is in some ways similar to the coarser ring of pseudopods of *Pedinella*, the functioning of the flagellum is quite different (Fig. 5.11a). Sponges are generally assumed to have been derived from collar–flagellate protozoans, some of which are colonial but do not display the cellular differentiation characteristic of sponges. Some planktonic marine choanoflagellates make complex basket-like loricas from precisely arranged siliceous rods[148] (Fig. 5.12).

Two groups of small flagellates with internal silica skeletons, the *Silicoflagellida* and Ebriida, classified as separate orders by some zoologists, may be offshoots of the Chrysophyceae, but a preliminary report that a silicoflagellate contains chlorophyll *c* suggests that this group may have a closer relationship with other chromophytes, possibly Bacillariophyceae. *Dictyocha* (Fig. 5.10k) is a common pigmented planktonic silicoflagellate whose typical annular skeleton, which is at least partly extracellular, is made of tubular elements; a single anterior flagellum and slender pseudopods emerge from the margins of the body, whose surface membrane is highly convoluted, with thin strands interconnecting islands of cytoplasm containing numerous chloroplasts and dictyosomes.[270] The colourless *Ebriida* are even less well known, but *Ebria* and *Hermesinum* (Fig. 5.10l) have a spicular internal skeleton of solid rods and swim by means of two flagella, thought to be held one in front and one behind. Many forms of skeleton belonging to both of these groups are found as fossils, mostly from tertiary and quaternary deposits, although cretaceous silico-flagellates are known.

Members of the *Xanthophyceae* (e.g. *Heterochloris*) appear similar to the Chrysophyceae in the possession of one long anterior flagellum with two rows of hairs and one short smooth flagellum, but differ in the colour of the chloroplasts, which are green or yellow–green, and in the food store, which is mainly lipid. Among the few unicellular species are amoeboid and coccoid forms as well as flagellates, and the latter appear similar to the zoospores of the more abundant filamentous and thalloid species e.g. *Bumilleria*[175] (Fig. 5.10m). Colourless species are rare, but the simpler fungi are thought to be related to the filamentous Xanthophyceae, which could account for the similar flagellation of fungal zoospores and these algae. The zoospores of filamentous members of the *Eustigmatophyceae*, e.g. *Vischeria* (Fig. 5.10n), are similar to the free flagellates of this group;

Fig. 5.13 Some representative members of the Dinophyceae; a, *Gymnodinium amphora* (30 μm) : b, *Peridinium tabulatum* (45 μm) : c, *Exuviaella marina* (40 μm) : d, *Polykrikos schwartzi* (150 μm) : e, nematocyst of *Polykrikos* (12 μm) : f, *Oxyrrhis marina* (20 μm) : g, *Ornithocercus splendidus* (120 μm) : h, *Craspedotella pileolus* (100 μm) : i, *Cymbodinium elegans* (500 μm) : j, *Kofoidinium pavillardi* (500 μm) : k, *Noctiluca scintillans* (800 μm) : l, *Oodinium ocellatum* (60 μm) : m, *Haplozoon clymenellae* (250 μm).

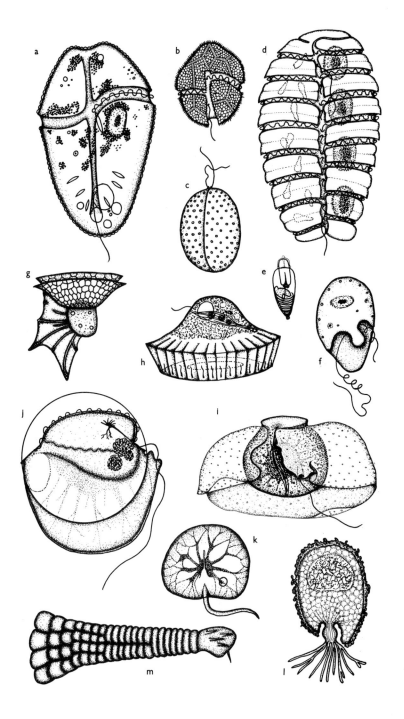

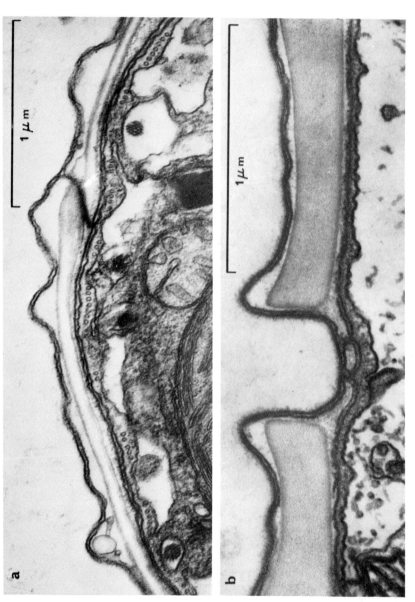

Fig. 5.14 Electron micrographs illustrating the structure of the theca as seen in sections of the dinoflagellates *Glenodinium foliaceum* (a) and *Ceratium hirudinella* (b). Beneath the cell membrane are membranous alveoli, which in these examples contain thin (a) and thick (b) skeletal plates, and groups of microtubules occur at the inner side of the pellicular layer. Micrographs by J. D. Dodge and R. M. Crawford.[64]

these species have recently been separated from the Xanthophyceae on the basis of the single flagellum, fine structural characters of the chloroplast, the different position of the eyespot and the absence of a prominent golgi body in the flagellate—they also have different xanthophyll pigments.[103] Members of the **Rhaphidiophyceae**, e.g. *Goniostomum* (Fig. 5.100), with numerous chloroplasts and trichocysts, seem to share a number of features with the Xanthophyceae, but species of this small group have not been extensively studied; the structure and origin of flagella hairs in *Vacuolaria* has recently been described.[102a]

Three groups of algae, the Phaeophyceae, the Bacillariophyceae and the Dinophyceae have all been found to contain chlorophylls *a* and *c*, but to lack chlorophyll *b*. These groups are quite different from one another although they are all brown. The diatoms with their yellow–brown chloroplasts and siliceous cell walls and particularly the filamentous or thalloid brown algae cannot be considered of direct concern to protozoologists and will not receive further discussion here. By contrast the **Dinoflagellida** (= Dinophyceae) contain an abundance of pigmented and colourless flagellate forms, some of which are colonial. Many of these flagellates are much modified in structure, including both autotrophic and heterotrophic species, and others have life cycles that are complex or involve special relations with animals as zooxanthellae (p. 280) or as parasites. There are also a few coccoid and filamentous autotrophic forms.

The nuclei of dinoflagellates are thought to be primitive since they share some features with the nuclear apparatus of procaryote cells—the absence of basic proteins in the nucleus and the construction of chromosomes from many fine fibrils[58] (Fig. 2.19, p. 45); in addition the chromosomes remain condensed in interphase and lack centromeres. These nuclei also show eucaryote features since they possess many chromosomes enclosed by a nuclear membrane and undergo a form of mitosis (p. 52). Other organelles of the dinoflagellates, by contrast, show at least as much complexity as in any other chromophyte group; the chloroplasts, flagella, cellulose theca and other special organelles are all highly developed.

The most characteristic features of a dinoflagellate may be illustrated by such common forms as *Peridinium* or *Gymnodinium* (Fig. 5.13a, b). The sculptured theca is grooved equatorially to form a girdle occupied by the transverse flagellum and a second groove passes downwards from the girdle at one side and forms the sulcus from which the longitudinal flagellum emerges. The form and movement of these two flagella are described on pp. 35, 36 and 115. The theca or pellicle is often thickened, and, in addition to the plate-like appearance visible in the light microscope, which is due to the presence of deposits of cellulose within membranous alveoli (Fig. 5.14), there is an ultrastructural pattern which appears to be related to the arrangement of underlying micro-

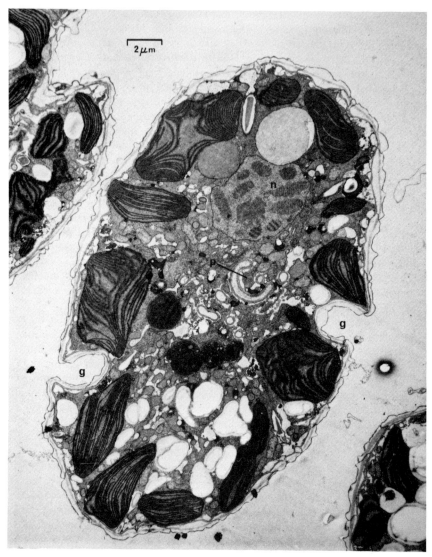

Fig. 5.15 Electron micrograph of a section through the dinoflagellate *Woloszyn-skia coronata* showing the position and appearance of the pusule system (p), the nucleus (n), the pellicular alveoli of the theca and their infolding to form the girdle (g). Micrograph by J. D. Dodge and R. M. Crawford (unpublished).

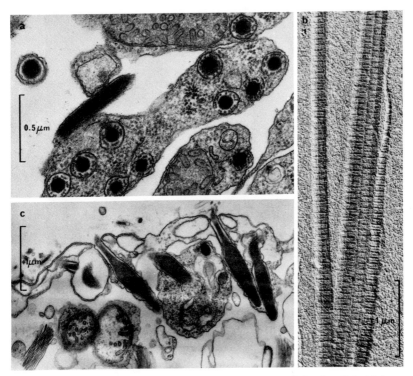

Fig. 5.16 Electron micrographs showing the structure of dinoflagellate tricho- cysts. Mature trichocysts of *Oxyrrhis marina* are shown in transverse (a) and longitudinal (c) sections,[66] and extruded trichocyst threads of *Exuviaella marie- lebouri* are shown in a shadowed preparation (b) (unpublished). Micrographs by J. D. Dodge and R. M. Crawford.

tubular fibrils.[64] The system of membranous alveoli and microtubules seen in the dinoflagellate pellicle is reminiscent of the pellicular structure of ciliates (p. 186), where however the cavity of the alveoli does not contain skeletal plates. Internally there are numerous yellow or brown chloroplasts and a large nucleus containing the permanently condensed chromosomes. A vacuole known as the pusule opens by a canal near the base of the transverse flagellum; its function is not known, but it seems likely that pusules are analogous to contractile vacuoles, although they are not known to contract in the manner expected of contractile vacuoles.[61] Sometimes the pusule forms an extensive system of vacuoles and canals in a specialized vacuolated region of the cytoplasm (Fig. 5.15). Beneath the theca numerous trichocysts are found which are comparable in structure to those of ciliates

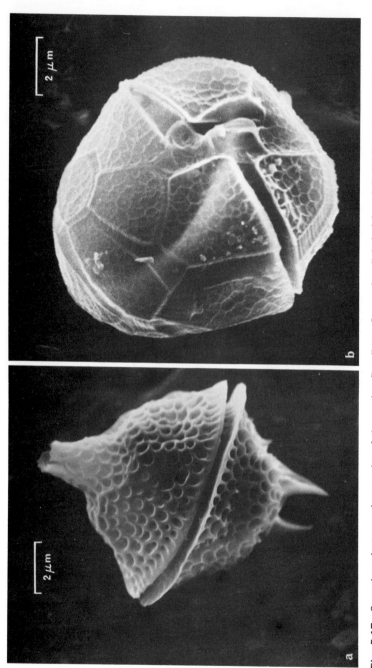

Fig. 5.17 Scanning electron micrographs of the marine flagellates *Gonyaulax digitale* (a) and *Peridinium cinctum* (b); the equatorial girdle is seen in both species and the sulcus in *Peridinium*. Micrographs by J. D. Dodge (unpublished).

(Fig. 5.16, c.f. Fig. 8.12, p. 201). Complex nematocysts that might easily be mistaken for coelenterate nematocysts are found in *Polykrikos* (Fig. 5.13e). Many dinoflagellates, both pigmented and colourless, ingest prey at an oral region near the posterior end of the sulcus.[65] Reproduction is normally by binary fission, which may be longitudinal or transverse, and in which parts of the pellicular theca may be passed on to each daughter, but multiple fission is characteristic of some species. Toxins produced by such dinoflagellates as *Gonyaulax* and *Gymnodinium* may be accumulated by filter-feeding molluscs and have caused deaths among people who have eaten the shell-fish.[190]

In a small number of species two similar flagella are borne at the

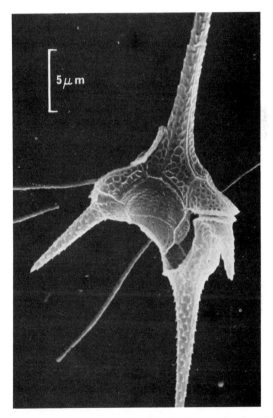

Fig. 5.18 Scanning electron micrograph of the freshwater dinoflagellate *Ceratium hirudinella* showing the thecal plates, the equatorial girdle and the large sulcus groove where food is taken into the body. Micrograph by J. D. Dodge and R. M. Crawford.[65]

anterior end of the body, but one of these extends in front while the other lies transversely around the base of the extended one. In similar forms the transverse flagellum is flattened, like the transverse flagellum of typical dinoflagellates, but the theca still lacks the characteristic transverse and longitudinal grooves. These flagellates form the sub-order Adenina, which is thought to include the more primitive dinoflagellates; *Exuviaella* (Fig. 5.13c) and *Prorocentrum* are common planktonic members of the group.

In *Dinophysis* and its relatives the bivalved theca is well-developed and the margins of the transverse and longitudinal grooves are drawn out into prominent flanges, which reach enormous proportions in *Ornithocercus* (Fig. 5.13g); the girdle is here carried towards the anterior end of the body. Several genera of these forms are abundant in the marine plankton. The theca is also well-developed in the peridinians, of which *Peridinium*, and *Gonyaulax* (Fig. 5.17) and *Ceratium* (Fig. 5.18) are probably the best known of a number of genera that are very important in marine and fresh-water plankton. In this case the thick theca is composed of many plates and the body may be extended into spines, notably in *Ceratium*.[65] In spite of their thick thecae many of these species practise phagotrophy and some are colourless; food is caught by *Ceratium* using posterior pseudopodia and is taken in between flexible plates in the posterior groove area (Fig. 5.18).

Dinoflagellates of the *Gymnodinium* type (Fig. 5.13a) have a thin flexible theca, and appear to be basic to a number of lines of evolution, several of which lead to groups that are purely heterotrophic. One interesting line is that which includes *Polykrikos*, species of which have replicated the flagellar system so that in *P. schwartzi* (Fig. 5.13d) there are 8 transverse girdles and 8 longitudinal flagella one above the other in an organism which contains four nuclei; many of these species are colourless and prey on Metazoa and Protozoa which may be caught with the aid of nematocysts. Another interesting group studied recently by J. and M. Cachon[27, 28] includes *Noctiluca* (Fig. 5.13k), a well-known source of marine luminescence, and some bizarre colourless forms including *Cymbodinium* (Fig. 5.13i), which resembles a small veliger larva, and *Craspedotella* (Fig. 5.13h), which is like a small medusa and progresses by contractions of the velum

Fig. 5.19 Some representative members of the Euglenophyceae; a, *Euglena gracilis* (50 μm), with nucleus (*n*), stigma (*s*), plastids (*p*) with central paramylon deposits, two flagella arising in flagellar pocket (*fp*), the short one terminating at about the level of the flagellar swelling on the long one, and the contractile vacuole (*cv*): b, *Entosiphon sulcatum* (20 μm): c, *Phacus pleuronectes* (80 μm): d, *Trachelomonas hispida* (30 μm): e, *Colacium vesiculosum* (20 μm) often epizoic on crustaceans; f, *Peranema trichophorum* (60 μm). Mainly based on information by Leedale.[153]

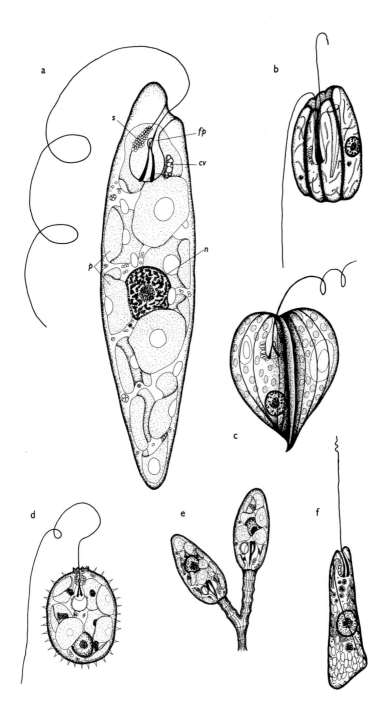

a

s

fp

cv

p

n

b

c

d

e

f

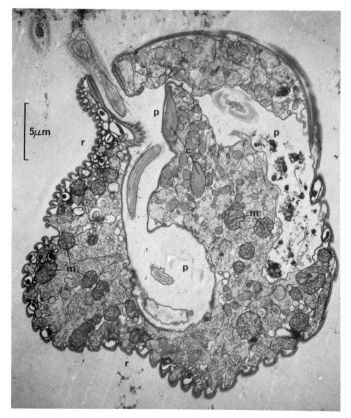

Fig. 5.20 Electron micrograph of an oblique transverse section through the anterior end of *Peranema trichophorum*. The surface of the body has numerous pellicular ridges (r) (see Fig. 5.22), and the two flagella emerge from the flagellar pocket (p), which is rather distorted in this specimen, so that several sections of the complex flagella are seen (cf. Fig. 5.21); alongside the flagellar pocket are the two pharyngeal rods (c) and numerous characteristic mitochondria are present (m). Micrograph by B. Nisbet (unpublished).

that surrounds the cytostome. These dorsoventrally flattened forms may be contrasted with the extremely vacuolated, spherical *Noctiluca* (Fig. 5.13k) and the laterally flattened *Kofoidinium* (Fig. 5.13j), a remarkable little animal that swims by two flagella and carries on top a saucer-shaped transparent shell which is held only by finger-like tentacles at two opposite sides of the shell margin. The products of multiple division of this organism look very much like *Gymnodinium* and pass through stages reminiscent of other dinoflagellates during their development; at any stage in the

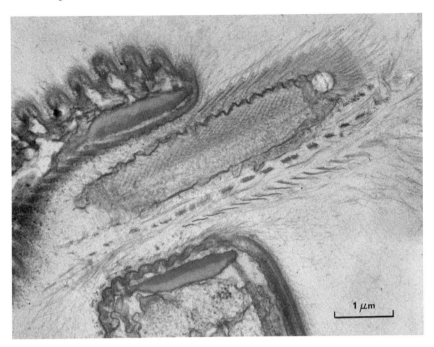

Fig. 5.21 Electron micrograph of a section of the flagellum of *Peranema* at the entrance to the flagellar pocket; the flagellum is complex, being surrounded by a variety of hairy projections and containing alongside the axoneme an intraflagellar rod that can be seen to have a striated pattern which is also seen in similar rods in the dinoflagellate *Oxyrrhis* and the trypanosome *Blastocrithidia*. Micrograph by B. Nisbet (unpublished).

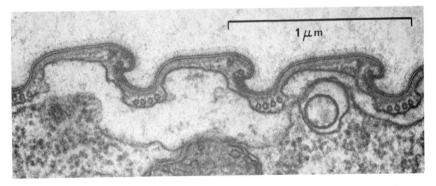

Fig. 5.22 Electron micrograph of a section of the pellicle of *Peranema* showing the ridges formed by dense material immediately beneath the cell membrane, and incorporating groups of microtubules. Micrograph by B. Nisbet (unpublished).

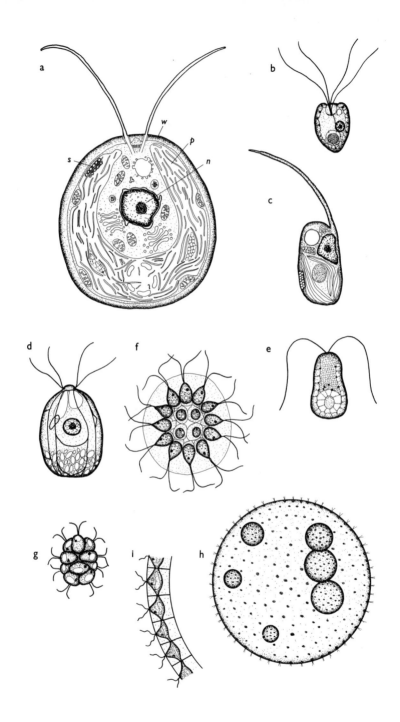

life cycle they may also undergo binary fission. The colourless *Noctiluca*, which may grow to about 1 mm in diameter, carries a long motile tentacle, near the base of which the one remaining (longitudinal) flagellum emerges; it is a voracious predator and undergoes multiple fission to produce uni-flagellate 'spores' which have the remains of a girdle. In *Oxyrrhis*, a colourless dinoflagellate which is common in infusion cultures from marine sources, there are two flagella with complex root structure, but the girdle is poorly developed[66] (Fig. 5.13f).

Finally there is a group of dinoflagellates which are parasitic in animals and plants, and whose relationships are betrayed by the production of

Fig. 5.24 Electron micrograph of scales from the flagellate *Pyramimonas* (Prasinophyceae). Micrograph by R. M. Crawford (unpublished).

numerous biflagellate zoospores, generally with a shape like *Gymnodinium*, and sometimes carrying chloroplasts. These are parasites of Protozoa, coelenterates, molluscs, worms, crustaceans, tunicates and fish, which live either as ectoparasites, feeding by rootlets entering the tissues of the host, e.g. *Oodinium* (Fig. 5.13l), or as endoparasites, some of which are multicellular, e.g. *Blastodinium* and *Haplozoon* (Fig. 5.13m).

Although the **Euglenophyceae** contain chlorophyll *b*, and are therefore

Fig. 5.23 Some representative members of the Chlorophyceae, Loxophyceae and Prasinophyceae; a, *Chlamydomonas reinhardi* (10 μm), with cell wall (*w*), plastid (*p*), stigma (*s*), and nucleus (*n*) in the central cytoplasmic region: b, *Pyramimonas tetrarhynchus* (25 μm): c, *Pedinomonas minor* (6 μm): d, *Polytomella caeca* (15 μm): e, *Dunaliella salina* (12 μm): f, *Gonium pectorale* (cells 12 μm): g, *Pandorina morum* (cells 15 μm): h, *Volvox globator* (colony 500 μm): i, arrangement of cells in the gelatinous wall of a *Volvox* colony.

classed as Chlorophyta, in most other respects they resemble various chromophytes more than the other chlorophytes. Members of the group show a number of characteristic features, which are well illustrated by *Euglena*[26] (Fig. 5.19a). Near the anterior end is an invagination, into which the contractile vacuole opens, and from the walls of which two flagella

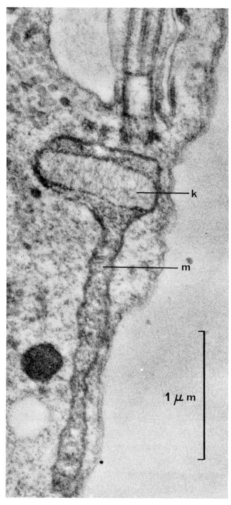

Fig. 5.25 Electron micrograph of a section through the trypanosome *Blasto-crithidia* showing the insertion of the flagellum at the base of the flagellar pocket and the adjacent kinetoplast (k) which is continuous with the mitochondrion (m). Micrograph by K. Vickerman.[220]

normally emerge (Figs. 5.6f, 5.20); in some species only one flagellum extends out of the flagellar pocket, and when there are two long flagella one of these is usually directed backwards, e.g. *Peranema, Entosiphon* (Fig. 5.19b, f). The intraflagellar rod is especially well developed in *Peranema* (Fig. 5.21), whose anterior flagellum also carries a diversity of extraflagellar structures that are better seen in transverse sections in Fig. 5.20. The main, anteriorly directed flagellum often has a swollen basal region which is presumed to have a photoreceptive function, since the pigmented stigma lies close-by. The carbohydrate storage product is paramylon (a β-1:3 glucan), which is deposited in crystalline granules of specific shapes. The pellicle is flexible in many species, and consists of interlocking longitudinal proteinaceous strips beneath the cell membrane, usually associated with microtubular fibrils[152] (Fig. 5.22). The nuclear membrane and the nucleolus (endosome) persist during mitosis, and, although an intranuclear spindle is present, the chromosomes lack centromeres (p. 52); some relationship with dinoflagellates may be inferred from these primitive nuclear features and the flagellar structure.

Green and colourless species of the Euglenophyceae occur in freshwater and seawater. They tend to be more abundant in smaller bodies of water with a high organic content, but some representatives of the group will almost always be found, especially in freshwater habitats. Most forms swim by flagella, but some move by peristaltic movements of the body; *Colacium* (Fig. 5.19e) is a stalked species that is often epizoic on crustaceans, *Trachelomonas* (Fig. 5.19d) is thecate and there are also loricate species. The nutrition of members of the group is diverse (p. 58); some of the heterotrophic species are unspecialized osmotrophs, while the phagotrophs have a well-developed cytostome which opens near the flagellar pocket and is often provided with stiff rods or tubes for ingestion (p. 67). A few parasitic species are known from invertebrates and amphibian tadpoles. A good recent account of the group has been written by G. F. Leedale,[153] and there are several books devoted to the genus *Euglena*.[26, 88, 291]

The other chlorophyte algae comprise three groups, of which the **Loxophyceae** and **Prasinophyceae** are small classes of flagellate forms which may be differentiated on the basis of flagellar characters[169] (Fig. 5.6), notably the pronounced long narrow flagellar tip of Loxophyceae, which are usually uniflagellate (Fig. 5.23c), and the flagellar scales and caducous hairs of Prasinophyceae, which usually have 2, 4 or more flagella, e.g. *Pyramimonas* (Figs. 5.23b, 5.24). The status of the Loxophyceae as a taxonomic group requires confirmation. The **Chlorophyceae**, by contrast, form a very large group with coccoid, filamentous, thalloid and colonial members as well as solitary flagellates. *Chlamydomonas* (Fig. 5.23a) is often taken as an example of flagellate members of the group, but

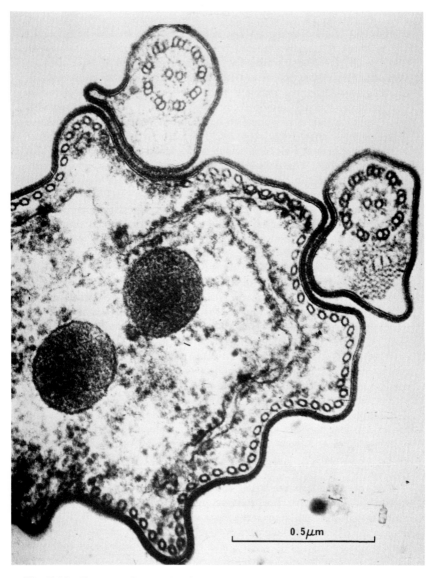

Fig. 5.26 Electron micrograph of a section through the bloodstream form of *Trypanosoma evansi* during the process of division. The two flagella are closely adherent to the body and a row of pellicular microtubules underlies the cell membrane which carries a thick surface coating in this stage of the life cycle. Micrograph by K. Vickerman.[273]

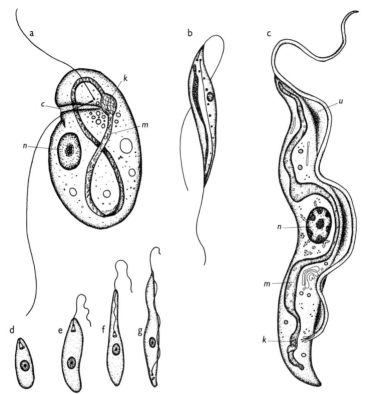

Fig. 5.27 Some representative members of the Kinetoplastida; a, *Bodo saltans* (10 μm), with nucleus (*n*) kinetoplast (*k*), mitochondrion (*m*) and cytostome (*c*): b, *Cryptobia helicis* (20 μm), from the seminal receptacle of the snail *Helix*: c, *Trypanosoma brucei* (20 μm), showing features of the bloodstream form including nucleus (*n*), kinetoplast (*k*), mitochondrion (*m*) and 'undulating membrane' (*u*): d, amastigote (leishmanial) form of trypanosome: e, promastigote (leptomonad) form of trypanosome: f, epimastigote (crithidial) form of trypanosome: g, trypomastigote (trypanosomal) form of trypanosome.

differences found in other genera include the presence of four flagella in *Carteria*, the lack of chlorophyll in the biflagellate *Polytoma* and quadriflagellate *Polytomella* (Fig. 5.23d) (though plastids are said to be present in *Polytoma*), and the lack of a thick cell wall in *Dunaliella* (Fig. 5.23e). Colonial aggregates of flagellate cells of the chlamydomonad type embedded in mucilage are found as flat plates in *Gonium* (Fig. 5.23f), as spherical clumps in *Pandorina* (Fig. 5.23g) and in *Volvox* as large hollow spheres in which thousands of cells are interconnected by protoplasmic strands (Fig. 5.23h, i). Both solitary and colonial forms are common in freshwater plankton. Similarities in the organization of the Chlorophyceae and higher green plants has led to the assumption that the green land plants evolved from these algae.

The **Kinetoplastida** contain two groups of heterotrophic flagellates, the biflagellate *Bodo* and its relatives, and the uniflagellate *Trypanosoma* and related forms; members of both of these groups are characterized by the possession of a DNA-rich organelle called the kinetoplast which is associated with the mitochondrion[197, 272] (Fig. 5.25). Several similarities suggest an origin of these flagellates near the euglenids or the dino-flagellates; in addition to the presence in all three groups of an intra-flagellar rod of comparable structure, unilateral flagellar hairs (Fig. 5.6d, e, f) and an extensive array of microtubules in the pellicle (Figs. 5.14, 5.22 and 5.26), the flagella (primitively two) arise from a pocket or 'reservoir' in all three groups, and food is taken in through an area near the flagellar pocket which may be reinforced with microtubules to form a cytostomial tube in some euglenids and kinetoplastids. Nuclear characters of the Kinetoplastida do not seem to be well enough known to make detailed comparisons possible, but there are similarities to euglenids since the nucleolus and nuclear membrane appear to persist through mitosis in *Bodo*, where there is an intranuclear spindle, and similar reports have been made for *Trypanosoma*.[275]

Species of *Bodo* (Fig. 5.27a) are very common in fresh and salt waters, particularly where these are polluted and abundant bacteria are present for food. The cytostome probably functions in pinocytosis, while bacteria are probably engulfed in the rostral region. Near the flagellar bases in bodonids is the rounded or discoid kinetoplast which has a fibrous central area and which connects at both sides with the two ends of the single long mitochondrion that forms a closed loop within the cell.[197] The two flagella emerge together near the anterior end and one extends forward while the other backwardly-directed flagellum lies free beside the body. In the related genus *Cryptobia* (Fig. 5.27b), species of which are parasitic in invertebrates and lower vertebrates, the backward flagellum is attached along the side of the body.

The trypanosomids are parasitic forms, often with polymorphic life cycles in which flagellate and non-flagellate stages may be present;[104]

Fig. 5.28   Some representative members of the Metamonadida; a, *Trichomonas muris* (15 μm) from large intestine of mice, showing nucleus (*n*), axostyle (*a*), costa (*c*), parabasal fibre (*p*) and undulating membrane (*u*): b, *Coronympha octonaria* (40 μm) from *Calotermes*: c, *Hexamita intestinalis* (10 μm) from large intestine of frogs: d, *Giardia muris* (10 μm) from small intestine of mice: e, *Joenia annectens* (100 μm) from *Calotermes*: f, *Saccinobaculus ambloaxostylus* (100 μm) from *Cryptocercus*: g, *Barbulanympha ufalula* (300 μm) from *Crypto-cercus*: h, *Trichonympha campanula* (200 μm) from *Zootermopsis*.

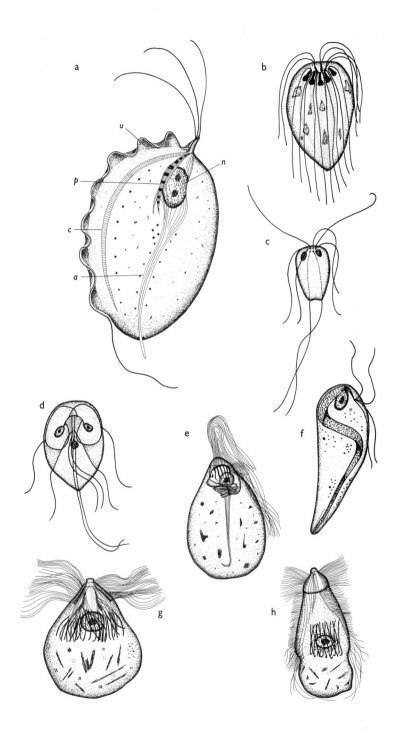

normally there is a single anteriorly-directed smooth flagellum accompanied by a second, barren, basal body.[272] *Leptomonas, Crithidia* and some related genera are found as gut parasites in insects; *Phytomonas* is found in plants, particularly latex-bearing species, and is carried from host to host by plant-sucking insects; *Leishmania* and *Trypanosoma* are parasites of vertebrates, including man, and are normally transmitted by blood-sucking invertebrates, usually insects or leeches. Four body forms are recognized, according to the position of the basal body and kinetoplast and the course followed by the flagellum; these are amastigote (=leishmanial), promastigote (=leptomonad), epimastigote (=crithidial) and

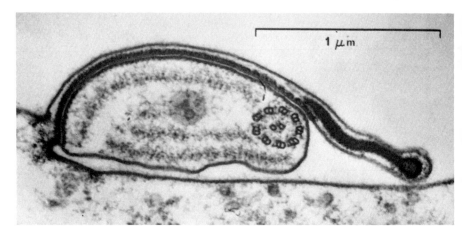

Fig. 5.29 Electron micrograph of a section through the recurrent flagellum and undulating membrane of *Trichomonas termopsidis*. The fold of the body surface that forms the undulating membrane is closely attached to the expanded flagellar membrane for a large part of its width and is reinforced at the margin by a dense lamella. Micrograph by A. Hollande and J. Valentin.[108]

trypomastigote (=trypanosomal) (Fig. 5.27d–g). They were previously called by the names given in parentheses, but this led to considerable confusion since, for example, *Leishmania* has a leptomonad stage in the life cycle as well as a leishmanial one, and all four body forms appear in the life cycle of some species of *Trypanosoma*. The main features of structure of the trypomastigote form of *Trypanosoma* are shown in Fig. 5.27c.

Species of *Leishmania* cause important human diseases, including the widespread kala-azar. The small amastigote parasites are found in the macrophages and lymphoid cells of infected vertebrates, where they grow and multiply. The parasites are transmitted by female sand flies (*Phlebo-*

tamus, Diptera), which ingest parasitized cells when they suck the blood of infected hosts. Within the fly the parasites leave the white blood cells, develop flagella to transform to the promastigote type, and multiply by

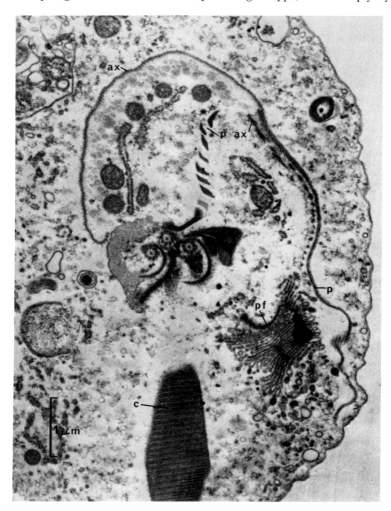

Fig. 5.30 Electron micrograph of a transverse section at the level of the basal bodies of *Trichomonas termopsidis* showing three basal bodies which have a close relationship with fibrous structures; these make connections at other levels with the costa (c), the parabasal fibre (pf) (with its associated dictyosomes) and through the pre-axostyle fibre (p ax) with the pelta (p) and the axostyle (ax). Micrograph by A. Hollande and J. Valentin.[108]

binary fission; flagellates migrate forward in the gut of the fly as far as the proboscis and are injected with saliva when the fly prepares to take another blood meal.

The life cycles of two species of *Trypanosoma* pathogenic in man will be described to illustrate different features of these organisms and of their relationships with their hosts. *Trypanosoma cruzi* is found in South and Central America in man and a diversity of mammals, between which it is transmitted by blood-sucking bugs, usually *Rhodnius*. Within the insect the ingested trypomastigote forms transform into epimastigotes which multiply repeatedly and eventually develop into small trypomastigote forms (metacyclic forms or metatrypanosomes) in the hind gut of the bug. The vertebrate is infected by contaminative means, through faeces of the insect being rubbed into skin lesions or unprotected mucous membranes, or in a number of other ways. The trypanosome enters lymphoid, macrophage and muscle cells of the vertebrate and there multiplies as amastigote forms. These may subsequently transform through promastigote and epimastigote stages to form trypomastigotes which circulate in the blood before re-entering the host cells and commencing a new amastigote multiplicative phase. The human disease caused by this parasite, Chagas' disease, is often fatal in children, and chronic in adults.

The African human trypanosomiasis, known as sleeping sickness, is caused by *T. brucei*, of which two recognized sub-species *T. brucei gambiense* and *T. brucei rhodesiense* are found respectively in western and eastern parts of tropical Africa. The vectors for these trypanosomes are species of the blood-sucking tsetse fly, *Glossina*. Within the fly the trypomastigote forms multiply in the midgut region and undergo a migration forwards to the salivary gland via the proboscis. In the salivary glands epimastigote forms multiply and metamorphose to form infective metacyclic trypomastigotes,[271] which may be injected into a new host at the next feed. Injected trypomastigotes circulate in the blood of the vertebrate and divide by binary fission; they do not have intracellular stages, but may later invade the cerebrospinal fluid, and fatally damage the brain. Other species of this type are important pathogens of domestic animals, particularly in Africa.

The trichomonad flagellates appear to be the central group among the **metamonad flagellates**, and provide a good introduction to the characteristic structures of these complex organisms. In *Trichomonas*[108] (Fig. 5.28a) three to five free anterior flagella and one recurrent flagellum arise together from a group of basal bodies. The recurrent flagellum is associated with a reinforced fold of the body surface, forming an undulating membrane (Fig. 5.29). A collar-like ridge around the bases of the flagella is supported by a sheet of microtubular fibrils called the pelta. The posterior end of each fibril of the pelta overlaps for a short distance with the anterior ends of microtubules which form the wall of the axostyle, and at about the same level there are the terminations of the filamentous pre-axostylar fibrils which connect the basal body complex with the axostyle wall (Fig. 5.30). The axostyle extends backwards, almost enclosing the nucleus, and may project at the posterior end of the body. Two large striated fibres arise from the basal body complex. The costa is a contractile fibre constructed of pleated lamellae which underlies the undulating membrane throughout its length. The parabasal fibre is ribbon-shaped and runs alongside the parabasal body, a very large array of golgi vesicles with periodic dense zones (Fig. 2.2, p. 17). Finally the basal bodies are the origin of a short rod which functions as an atractophore in the formation of the mitotic spindle (p. 51).

Other metamonads show multiplication of body organelles. This may take place by the multiplication of the caryomastigont structure—in *Coronympha octonaria* (Fig. 5.28b) there are eight flagellar groups, each associated with an axostyle, a nucleus and parabasal structures, and in *Metacoronympha* up to several hundred caryomastigonts are spirally arranged at the anterior end of the body. In the diplomonad flagellates *Hexamita* and *Giardia* (Figs. 5.28c, d) there are two nuclei and two sets of four flagella, but the other characteristic organelles of trichomonads do not appear to be present, so that these flagellates may have evolved from some other line of flagellates, although they could be reduced trichomonads. The parabasal structures are lost in the polymastigote flagellates *Oxymonas* and *Saccinobaculus* (Fig. 5.28e), where the axostyle is formed by 100–5000 microtubules in a compact ribbon, rather than an open cylinder, and has become the main motile organelle of the body;[95, 176a] replication of this form of caryomastigont is also found in other genera. In recent studies of some hypermastigote flagellates by A. Hollande and J. Valentin it was found that in *Joenia*[110] (Fig. 5.28f) a caryomastigont of the trichomonad type is found at one side of a large field of flagella; the three or four 'privileged basal bodies' of the special caryomastigont are associated with the axostyle, the parabasal apparatus and the atractophore, and at division all cell organelles except the group of privileged basal bodies are resorbed and reformed. Similar relationships between a group of privileged basal bodies and the axostyle, parabasal filaments and atractophore are found in each lobe of the bilobed *Barbulanympha*[107] (Fig. 5.28g), and four lobed *Staurojoenina*,[109] although in both of these there is a single nucleus and the cell organelles are not renewed at division. In *Trichonympha* (Fig. 5.28h) it appears that all that remains of the group of privileged basal bodies is a single basal body near the apex, but this still plays an important role in the formation of new organelles at division; a number of parabasal fibres arise in connection with the anterior flagella and are associated with golgi bodies posteriorly. In *Lophomonas* an axial calyx structure formed mainly from microtubules is dilated anteriorly to surround the nucleus and flagellar bases;[20] members of this genus are found in cockroaches and appear to have evolved independently of other hypermastigotes.

The majority of these metamonads are symbionts. A few of them are free living, e.g. *Hexamita*, some species of which may be found in polluted water, while other species are found in the digestive tracts of fish and amphibians. *Giardia* is found in vertebrates, including man, although it is not a serious pathogen. Species of *Trichomonas* are found in man and domestic animals; *T. foetus* is an important species which is sometimes responsible for abortion in cattle, but infection of the human genital tract by *T. vaginalis* is less serious. Almost all of the more complex metamonad flagellates are symbionts in insects, principally termites and wood-eating roaches. While the simpler forms often feed on bacteria or are osmotrophs, most of the larger forms engulf fragments of wood which are digested within food vacuoles. The intensive studies of Cleveland and others have shown the metamonads to be a most interesting group (see pp. 79 and 278), although only some examples from the diversity of complex forms have been mentioned here.

6

The Opalinid Protozoa

Opalinid Protozoa were found in frog faeces by Leeuwenhoek in the early days of microscopy. Superficially these organisms appear to be ciliates, and were once classified as such, but their morphology, form of division and life cycle are different from those characteristic of ciliates.[37, 286]

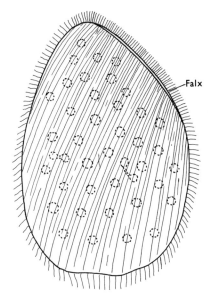

Fig. 6.1 A general view of *Opalina ranarum* showing the pattern of ciliary rows, the numerous nuclei and the falx along the right anterior margin of the body.

Full grown opalinids are found in the rectum of frogs and toads. *Opalina ranarum* from *Rana temporaria* is a leaf-shaped organism, from 300 to 800 μm long, about 300 μm wide and only about 30 μm thick, Fig. 6.1. There is some evidence that the animal is laterally flattened, and the side on which it usually lies on a microscope slide is the left side. Both surfaces of the body are densely covered with cilia, which are about 15 μm long and are borne in rows extending backwards for various distances from the right anterior margin of the body. These cilia move in close waves which can be seen sweeping back over the surface of the body as the animal swims forwards (Fig. 6.2). The direction of swimming is

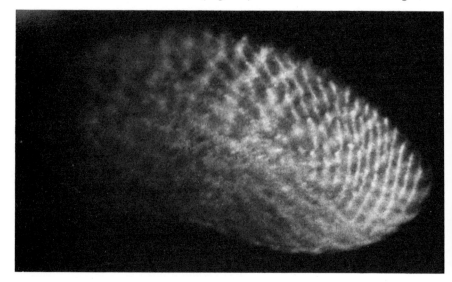

Fig. 6.2 Photomicrograph of *Opalina* taken under dark-ground illumination and showing the appearance of the metachronal waves on dorsal and ventral surfaces of the body.

changed by alteration in the direction of beat of the cilia, which coincides with the direction of movement of the waves and is related to the potential difference across the surface membrane (p. 43).

Between the rows of cilia the pellicle is thrown into narrow longitudinal folds, each supported by a single longitudinal ribbon of 12–24 microtubular fibrils[188, 287] (Fig. 6.3a). About six of these folds occur between adjacent ciliary rows, and in the troughs between the folds there are signs of pinocytotic activity; *Opalina* has no cytostome, and appears to feed by pinocytosis over much of the body surface. A short striated fibril passes forwards on the animal's left side from each kinetosome, extending only as far as the adjacent kinetosome, with which it makes contact (Fig. 6.3b); this fibril is not homologous with any of the fibrils found around the kinetosomes of ciliates. At the anterior margin of the body is a region called the falx (Fig. 6.1), which is a zone of cilia one cilium wide at the ends and 4 to 6 cilia wide for most of its length. The other ciliary rows arise at either side of the falx, which appears to be a region of some morphogenetic importance. At least some of the microtubules of the pellicular folds appear to be connected to the falcular kinetosomes.[287]

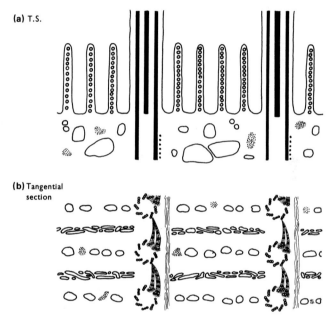

(a) T.S.

(b) Tangential section

Fig. 6.3 The structure of the pellicle of *Opalina* as seen (a) in transverse section and (b) in a tangential section grazing the surface at the level of the ciliary basal bodies. Regularly arranged groups of vesicles and other inclusions occur between the rows of cilia. Based on electron micrographs by C. Noirot-Timothée[188] and H. Wessenberg.[287]

Large numbers of similar nuclei are present in *Opalina*, although in other genera, e.g. *Zelleriella*, there may be only two nuclei. In *Opalina* no temporal relationship is observed between nuclear division and cell division, and there is no synchrony of division of nuclei, so that at any time a variable number of nuclei from none to many may be involved in mitosis. Both the nuclear membrane and the nucleoli appear to persist throughout mitosis. Division of the body is usually preceded by an increase in the length of the falx and in the number of kinetosomes in this region; then a cleft appears at the anterior end of the body and extends backwards across the body between the rows of cilia to divide the animal into two daughters whose sizes may be similar or quite unequal (Fig. 6.4). Occasionally the plane of division may cross the ciliary rows, and a new zone of falcular kinetosomes is developed in the posterior daughter if it has no part of the original falx.

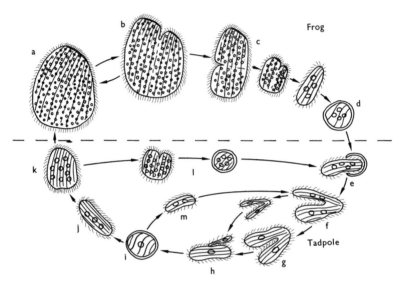

Fig. 6.4 The life cycle of *Opalina*; *a–b*, binary fission in the adult frog: *c*, repeated fission prior to formation of a cyst *d* which is voided by the frog: the cyst hatches *e* when eaten by a tadpole and further division *f, g* results in the formation of uninucleate gametes which fuse to form a zygote *h*: which encysts *i*: when this cyst is eaten by another tadpole it may grow into a normal trophic form *j, k, a*, or the small trophic form *k* may undergo division and asexual cyst formation *l*, or the form that hatches from the zygocyst may develop directly into a gamont *m*. Information by H. Wessenberg.[287]

These trophic forms of *Opalina* may be found throughout most of the year in the adult host, but during the breeding season of the amphibian the protozoans divide by a rapid series of longitudinal and transverse divisions to produce many small daughters with few nuclei and few ciliary rows. It is believed that hormonal secretions associated with control of the reproductive activities of the host are responsible for the onset of rapid division of the opalinids. The small daughters with about 3 to 6 nuclei form approximately spherical cysts about 30 μm across in the faecal mass of the host, and are passed out of the amphibian. The cysts survive for some weeks in water, but if they are eaten by tadpoles they quickly hatch in the gut. The protozoans which emerge from the cyst are gamonts which multiply by a series of longitudinal fissions to produce uninucleate fusiform gametes; among which slender microgametes and fatter macrogametes may be recognized; it is assumed that meiosis occurs during these divisions. Abundant gametes may be found in the hind-gut of the tadpoles, and fertilization takes place there by attachment of a microgamete to a macrogamete and the subsequent absorption of the former by the latter. Shortly after syngamy the zygote becomes rounded, secretes a cyst wall, and is seen to be uninucleate. The zygocysts are shed into the water with the faeces of the tadpole, and on ingestion by another tadpole they may hatch to produce another generation of gamonts and gametes, or may develop into multinucleate trophic forms, which are at first cylindrical and later become flattened; the fate of the excysting protozoan is apparently determined by the stage of development of the tadpole which has eaten it. Another generation of asexual cysts may be produced in the tadpole before metamorphosis and departure of the amphibian from the water, so there is ample opportunity for infection of all tadpoles in the vicinity. The life cycle of *Opalina* is illustrated in Fig. 6.4.

Opalinids differ from typical ciliates in the possession of many nuclei of only one type, the occurrence of a division plane running between ciliary rows, the form of the infra-ciliature, and the sexual process of syngamy. The multinucleate condition, the occurrence of a division plane across the ciliary rows in some fissions, the lack of centrioles in nuclear division and the dependence on kinetosomes rather than centrioles in morphogenesis following division are not generally characteristic of flagellates. It seems likely that ciliates and opalinids evolved along somewhat divergent lines from the same or similar ancestral flagellates; the opalinids are now distinctly separate from both of the other groups, although they exhibit some flagellate features and some ciliate features. There is no evidence that any opalinid harms its host at any stage in the life cycle, although it may derive nourishment from mucous secretions as well as from gut fluids in the rectum.

7

The Amoeboid Protozoa

The sarcodine Protozoa[126] normally possess pseudopodia, most forms of which are locomotor organelles of a particularly flexible type. While examples from several amoeboid groups may secrete tests, shells or skeletons, these do not form a complete covering for the body, and naked pseudopodia are used for locomotion and for the collection of food. The cell surface layers are not differentiated in special ways as they often are in flagellates, ciliates and sporozoans, and perhaps this lack of cortical differentiation is a result of the cytoplasmic mobility necessary for pseudopodial movement. In spite of their lack of cortical organization, many sarcodines show some remarkably complex structures, particularly in external shells and in internal skeletons, both living and non-living. Major features of interest in the Sarcodina centre around the pseudopodia and their use in locomotion, the composition, organization and construction of the shells and skeletons, the specialized life cycles and reproduction of members of some groups, and relations with other organisms in various symbiotic relationships.

MOVEMENT OF AMOEBOID PROTOZOA

Amoeboid movement is characterized by the flowing of cytoplasm which is made visible under the light microscope in most examples by the movement of granules (including mitochondria, 'vacuoles' and crystalline structures). The forms of this movement are diverse, and the mechanisms proposed to account for them are if anything still more diverse. Some better known examples of movement by amoebae will be described, and brief mention will be made of the more widely quoted mechanisms that have been advanced to account for them. For a full treatment of this topic reference should be made to reviews of the subject.[5, 6, 123, 230]

Four major types of pseudopodium are recognized: the broad blunt lobopodium characteristic of *Amoeba* (Fig. 7.1a); the needle-shaped, slender filopodia which may be branched but do not form a reticulum (Fig. 7.1b); similar pseudopodia, normally unbranched, which are supported by some central 'skeletal' structure, and are called axopodia (Fig. 7.1c); and the slender repeatedly branching and anastomosing reticulopodia (Fig. 7.1d).

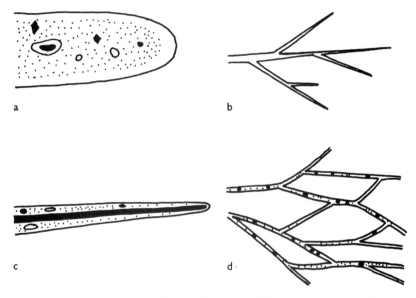

a   b

c   d

Fig. 7.1   Forms of pseudopodia found in amoeboid Protozoa; *a* a lobopodium, *b* filopodia, *c* an axopodium and *d* reticulopodia.

Pseudopodia of the lobose type seem to show the greatest diversity. In these forms there is a characteristic cytoplasmic interconversion between ectoplasmic gel and endoplasmic sol, normally with sol formation in areas of withdrawal and gel formation in areas of extension. The most familiar pseudopodia are those of the carnivorous amoebae *Amoeba proteus* and *Chaos carolinensis*, which normally have several more or less tubular pseudopodia, carrying a hyaline cap at the tip (Fig. 7.2a), and frequently containing granules of various sizes flowing in the cytoplasm. These amoebae feed by the formation of pseudopodial food cups. The extension of pseudopodia in locomotion is followed by a flowing of the

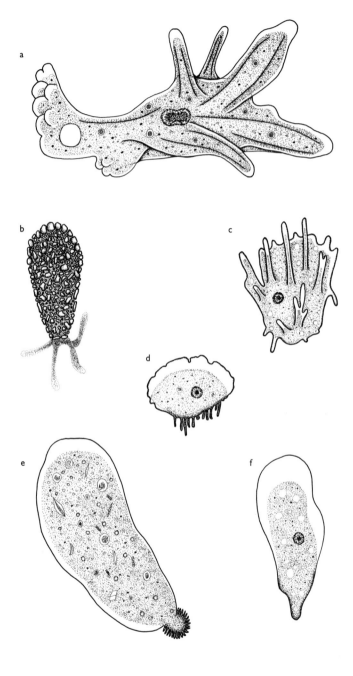

a

b

c

d

e

f

body cytoplasm, including surface gel that has been solated, into the ectoplasmic envelope of the advancing pseudopods. Forward locomotion occurs because of the occasional attachments of the pseudopodium to the substratum. For many years it was accepted that the forward flow of endoplasm is achieved by contraction of the rear end of the ectoplasmic tube,[123] but recently this explanation has been challenged. R. D. Allen has collected substantial evidence in support of a frontal contraction hypothesis to explain the movement of these amoebae;[5] according to this 'fountain-zone' hypothesis the source of motile forces is a contractile activity in the transformation of endoplasm to ectoplasm near the advancing tip of the pseudopods (see also p. 164).

Several tubular pseudopodia may also extend from the apertures of such testate (shelled) amoebae as *Arcella* (Fig. 7.10a, b) or *Difflugia* (Fig. 7.2b). These pseudopods extend in a similar manner to those of *Amoeba proteus*, but the locomotion is different because the shell must be carried or dragged along. An extended pseudopod of *Difflugia* forms attachments to the substratum and, following the appearance of birefringent strands within the pseudopod, it shortens and pulls the shell nearer to the attachment points.[290]

The giant herbivorous amoeba *Pelomyxa palustris* (Fig. 7.2e) does not form tubular pseudopods, but retains a slug-like or sac-like shape even when moving. In this type the hyaline-cap is at the rear of the amoeba, and feeding, which also occurs near the rear end, is said not to involve the formation of food cups. The forward flow of endoplasm in this amoeba is thought to result from the contraction of ectoplasm at the rear of the organism.

In many of the small amoebae, such as *Hartmanella* and *Naegleria* which are common in soil, and the parasitic form *Entamoeba* (Fig. 7.8), only one pseudopodial lobe is usually found at any one instant. This is not formed by a smooth flowing movement, but the whole lobe appears in a sudden burst, and no movement of the pseudopod occurs for a while before another eruption leads to the formation of a further pseudopodial lobe. Usually such pseudopodia are clear and lack granules; in small amoebae they are very thin. While pseudopod formation is eruptive, the main body of the amoeba flows more smoothly behind the advancing part of the organism. Such movements could be caused by contraction of the ectoplasm of the main body. Clear pseudopodia that are broad but very thin also characterize other, often larger, amoebae, such as

---

Fig. 7.2 Some representative examples of amoebae with lobopodia; *a, Amoeba proteus* (500 μm); *b, Difflugia oblonga* (250 μm); *c, Mayorella bulla* (100 μm); *d, Flabellula citata* (20 μm); *e, Pelomyxa palustris* (2 mm); *f, Vahlkampfia limax* (30 μm).

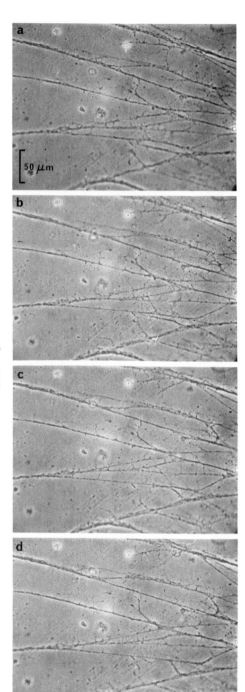

Fig. 7.3 Photomicrographs of part of the pseudopodial reticulum of a foraminiferan taken at 10-second intervals.

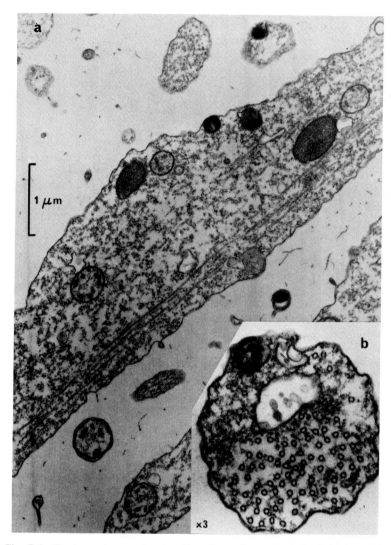

Fig. 7.4 Electron micrographs of sections through pseudopodia of the foraminiferan *Allogromia*. In (*a*) the pseudopodia are of various sizes and are sectioned at a variety of angles; they show longitudinal microtubules, occasional dense bodies and mitochondria with few cristae. In (*b*) the transverse section of a single pseudopod shows sectioned microtubules clearly. Micrograph by R. H. Hedley (unpublished).

Mayorella (Fig. 7.2c). Amoebae that rarely form pseudopodia often have a heavy ectoplasmic layer, e.g. *Thecamoeba*.

Fine *filopodia* are found on a few smaller amoebae and are usually rather transparent with few granules; they form small or extensive branching systems within which each tapering branch shows some independence of movement, although extension and retraction of all branches also occurs simultaneously. Filopodia are very similar to reticulopodia, but have fewer granules and show few or no anastosomes between the fine branches of the pseudopodia; these distinctions are not substantial, and further study may show them to be worthless.

Reticulopodia, best known from Foraminifera, are characterized by the network formed by very fine pseudopodial strands with more or less independent movement, along which a two-way flow of granules may be seen[125] (Fig. 7.3). In many parts of the network a number of strands run together so that some pseudopodial elements are thicker and have many streams of granules moving in each direction. These granules are probably mostly mitochondria, but such other inclusions as bacteria captured for food are also present. The diameter of the thinnest branches of the network is 1 μm or less, and within this thickness there appears to be a two-way flow of the cytoplasmic contents. Numbers of longitudinal microtubular fibrils occur within the pseudopods[176] (Fig. 7.4), but it is not yet clear whether the cytoplasm flows actively over these fibrils or whether forces which are generated between fibrils move the fibrils and also the surrounding cytoplasm; quite possibly both forms of active movement are involved, especially in extending or retracting pseudopods. Presumably the entire pseudopodial reticulum has contractile properties, for even heavy shells are moved about.

The *axopodia* of the heliozoans *Actinophrys* and *Actinosphaerium* (= *Echinosphaerium*) have been intensively studied in recent years. These fairly slender straight pseudopods have a central fibrous axis which extends within the body of the organism to the membrane of the nucleus or one of many nuclei (Fig. 7.5a). The fibrous axis is seen in transverse section as a double spiral of microtubular fibrils whose precise arrangement is maintained by cross-links between the fibrils[142a, 163, 215, 260] (Figs. 7.5b, 7.6). The axopodia are several hundred micrometres in length and may extend and shorten slowly, presumably by growth or dis-assembly of the microtubular fibrils of the axis.[283] Axopodia become attached to the surface over which a heliozoan is moving, and the slow retraction of anterior pseudopods and extension of posterior pseudopods moves the organism along by a combined gliding and rolling action (Fig. 7.5e, f). In addition to changes in axopod length, there is cytoplasmic flow up and down outside the fibrous axis in a number of longitudinal streams, as can be seen by the flow of granules and small food particles. Flagellates and ciliates

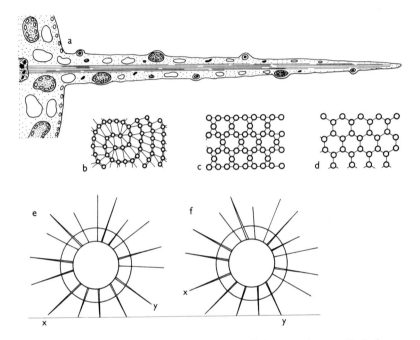

Fig. 7.5 The structure of various axopodia and the use of axopodia in loco-motion by the heliozoan *Actinosphaerium*. *a*, A diagram of a longitudinal section of an axopodium of *Actinosphaerium* showing the extent of the axial bundle of microtubules from the surface of a nucleus at the left to the tip of the axopod-ium. Diagrams showing the arrangement of microtubules seen in parts of cross sections of axopodia of *b*, a heliozoan; *c*, a centrohelid and *d*, a radiolarian may be compared with electron micrographs seen in Figs. 7.6, 7.7 and 7.21 respectively; in the Radiolaria Nassellaria the microtubules form a more open hexagonal array with 12-membered rings.[29] The relative lengths and positions of axopodia of an *Actinosphaerium* moving over a substrate are shown at *e* and at *f* 10 minutes later; the heliozoan is moving towards the right by a combination of rolling and gliding. Information for *e* and *f* from photomicrographs by C. Watters.[283]

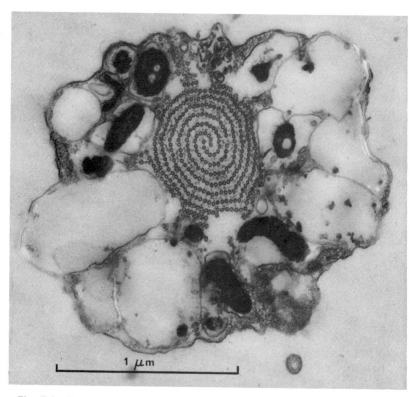

Fig. 7.6 Electron micrograph of a section through an axopodium of *Actino-sphaerium nucleofilum*. Note the two intercoiled spirals of microtubules and the vacuolated cytoplasm. Micrograph by A. C. Macdonald.

are caught by these heliozoans upon the adhesive axopodia which retract to carry the prey towards the body surface for enclosure in a food vacuole.

In some of the smaller heliozoans (centrohelids) the axopodia are more slender and have their origin near a short fibrous rod in the body rather than from the nuclear surface[14] (Fig. 7.7). The arrangement of the micro-tubular fibrils in a space-centred hexagonal array (Fig. 7.5c), and the more rapid contractility of the arms (remarkably fast in some species), are other differences of these axopodia from those of the larger Actinophryid helio-zoans described above. Slightly different arrangements of microtubules have been found in fibrous axopodia of radiolarians (Figs. 7.5d, 7.21), some of which show a very rapid contractility.[106] In general radiolarian pseudopods are less well known; in some cases they appear to be reticular networks; but more often they are slender filopodia or axopodia which may be

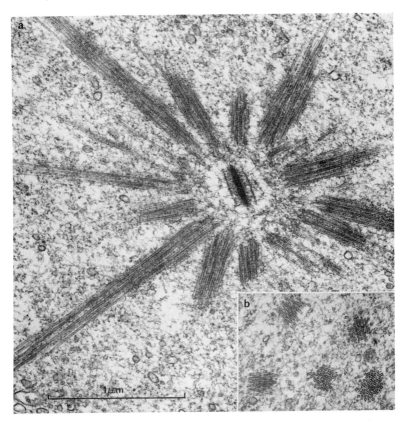

Fig. 7.7 Electron micrographs of sections of *Raphidiophrys ambigua* showing the bundles of microtubules which support the axopodia. In *a* the bundles of microtubules radiate around a 'centroplast' structure, and in *b* sections across several bundles are shown. Micrographs by C. F. Bardele.[14]

fibrous and contractile or supported by rigid siliceous spicules[29, 106] (Fig. 7.20).

All of these amoeboid forms show some form of naked pseudopodium within which movements take place with or without clear signs of contractile activity. The characteristics of these movements are diverse and no single explanation is likely to account for all forms of movement shown by amoebae; there are clearly several different patterns of evolutionary exploitation of the contractile properties of cytoplasmic components. While streaming movements around axopodia or within reticulopodia of foraminiferans may depend upon the axis of microtubular fibrils, contractile movements in other cases are probably due to interactions between filamentous microfibrils. Actin-like fibrils occur in *Acanthamoeba*,[203] and in *Amoeba proteus* labile thin actin-like filaments and more stable thicker filaments (probably myosin) are both present in the cytoplasm.[202] Only further research will show how these fibrils are involved in rear-contraction, fountain-zone contraction and other locomotory processes.

THE DIFFERENT FORMS OF AMOEBOID PROTOZOA[90,145]

The Rhizopoda

The naked amoebae with lobose pseudopodia are mainly distinguished from one another by the form of the pseudopodia and such factors as the number and form of the nuclei. Different species possess cylindrical, wedge-shaped, conical or webbed pseudopods, some of which can be seen in Fig. 7.2.

The cytoplasm is usually divisible into a clearer ectoplasmic region and a more granular endoplasm containing one or many nuclei and a variety of vacuoles and food storage granules. The free living members of this assemblage range in size between a few μm and a few mm, so that it is not surprising if such biological features as locomotion and feeding vary somewhat within the group. Among the larger forms are carnivores, e.g. *Amoeba* and *Chaos*, eating mainly ciliates and flagellates, and herbivores, e.g. *Pelomyxa palustris*, eating mainly diatoms, while the smaller forms, e.g. *Hartmanella*, feed mainly on bacteria. Most of the better-known amoebae are found in fresh waters, but some are marine and others live in soils. A particularly interesting small amoeba is *Naegleria*, which is common in soil and may encyst when dry or develop flagella when flooded (p. 261); it may be regarded as a form linking the amoebae with flagellates.

Parasitic lobose amoebae are generally small, usually between 5 and 50 μm, and are classified into genera according to the arrangement of chromatin material in the nucleus—it may be diffuse, central or peripheral. The most important amoeba parasitic in man is *Entamoeba histolytica* (Fig. 7.8), the cause of amoebic dysentery. Under certain circumstances the amoebae may break down the epithelium of the large intestine and cause ulceration and bleeding, following which the amoebae may spread through the portal vein to the liver and perhaps other organs and produce abscesses, rupture of which can be fatal. Food vacuoles normally contain bacteria and host-cell remains, but in dysentery cases they often contain erythrocytes and fragments of other tissue cells. The amoebae reproduce by binary fission and form thin-walled cysts by means of which the parasite is spread from one host to another. The cysts of different parasitic amoebae may be distinguished by the number of nuclei present; *E. histolytica* has four nuclei in the cyst, while *Entamoeba coli* which is also common in the human digestive tract has 8. Amoebae of several other genera are also common parasites of man but most of these are non-pathogenic, and are probably exclusively bacterial feeders. Occasionally such 'soil amoebae' as *Hartmanella* and *Naegleria* are found in association with human diseases; invasion of the central nervous system by such amoebae has proved fatal.

The **Myxomycetes** (Mycetozoia) are organisms with an initial amoeboid stage from which a multinucleate plasmodium or a multicellular colony develops and culminates in a sporangium of the type found in

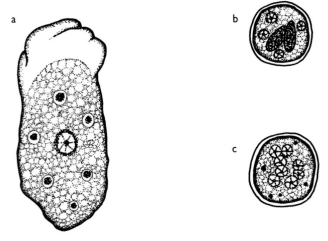

Fig. 7.8 *Entamoeba.* A trophic cell of *Entamoeba histolytica* is shown at *a* (10–20 μm long), and a cyst of *b, E. histolytica* with 4 nuclei (5–20 μm) is compared with a cyst of *c, E. coli* with 8 nuclei (10–30 μm). The characteristic form of the nucleus is shown and the shape of the 'chromatoid bodies' in the cyst of *E. histolytica* is shown at *b*.

many lower fungi.[174a] Three groups are frequently classed here. 1, The cellular slime moulds (Acrasida) in which large numbers of amoebae form multicellular aggregates that differentiate to form sporangia, e.g. *Dictyostelium*. 2, The free living plasmoidal slime moulds (Eumycetozoida) in which amoeboid and flagellate stages may precede the formation of a syncytial plasmodium; this takes the form of an extensive branching system of tubular channels that ramifies over soil, rotten wood or decaying leaves, and may form sclerotia or sporangia, e.g. *Physarum, Badhamia*. 3, The parasitic plasmodial slime moulds (Plasmodiophorida) which inhabit plant tissues and cause such diseases as club-root in cabbages, e.g. *Plasmodiophora*. There seems to be little justification for including the second and third groups with the Protozoa; although these organisms have amoeboid and flagellate stages, and the plasmodia of free living forms have been much studied as examples of cytoplasmic streaming, they will not be mentioned further in this book. The acrasian slime moulds on the other hand are more widely accepted as Protozoa, and have few fungal features.[22]

Several species of *Dictyostelium* are available as laboratory cultures; they are probably naturally coprophilous. The spores of *Dictyostelium* hatch in suitable media, and the amoebae which emerge consume bacteria, grow and divide. When the bacteria are plentiful the population of amoebae increases logarithmically, and the amoebae remain independent trophic organisms. In conditions of food shortage the amoebae aggregate, apparently under the influence of a substance called acrasin, which either is cyclic-AMP, or causes the release of this substance. At this time there is a very interesting change in character of the organisms from a state of mutual avoidance to a state of mutual adhesion. Aggregation involves the streaming together of vast numbers of amoebae into a large mass, which subsequently migrates as a 'slug', composed of separate amoebae moving over one another. After a period of migration the movement stops and amoebae towards the anterior end begin to rise up to form a stalk, with a differentiation of the cells and the production of cellulose. Other amoebae creep up the outside of the stalk to produce more stalk cells. Finally, the last amoebae to climb up the stalk differentiate to form spores, with resistant walls and a store of reserve food materials, which are seen as a spherical spore mass at the head of the stalk (Fig. 7.9). These organisms are currently attracting considerable interest in research, particularly from the point of view of the changes in biochemical characteristics associated with the aggregation phenomenon and with the differentiation of cells to form the stalk and spores;[186] this work will be facilitated by the recent isolation of *Dictyostelium* in axenic culture.

The **testaceous amoebae** of fresh water, damp soil and mossy habitats have pseudopods that are blunt and cylindrical, branched and pointed or even filopodial. They also vary in the nature of the shell, and shell characters are mainly used in classification. The shell of *Arcella* (Fig. 7.10a, b), common on weeds and mud in more or less stagnant fresh water, varies from pale yellow to brown, according to age, and is composed of layers of organic (tectinous) material secreted by the amoeba. In other forms, such foreign materials as sand grains and diatom frustules may be incorporated in the shell, e.g. *Difflugia* (Fig. 7.2b) and *Centropyxis* (Fig. 7.10c, d). The shell in *Euglypha* and related genera is formed from imbricated siliceous plates (Figs. 7.10e, f, 7.11)[101a]; these are preformed internally and fitted into precise positions during shell formation. The

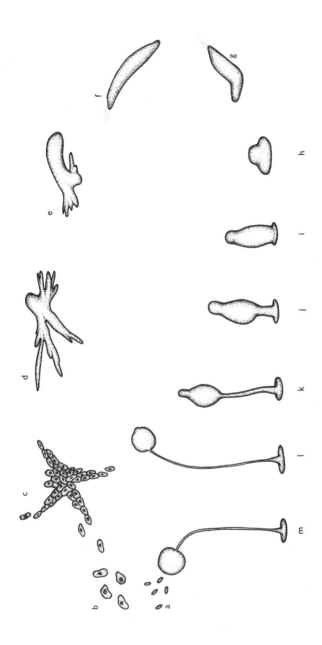

Fig. 7.9 The life cycle of the cellular slime mould *Dictyostelium discoideum*. Spores (*a*) released from sporangia hatch to release amoebae (5–15 μm) which grow and reproduce (*b*). Eventually the amoebae aggregate (*c*), and the resultant cell mass (1–2 mm) migrates (*d–g*) and later grows upwards to form a sporangium (*h–m*) (see text).

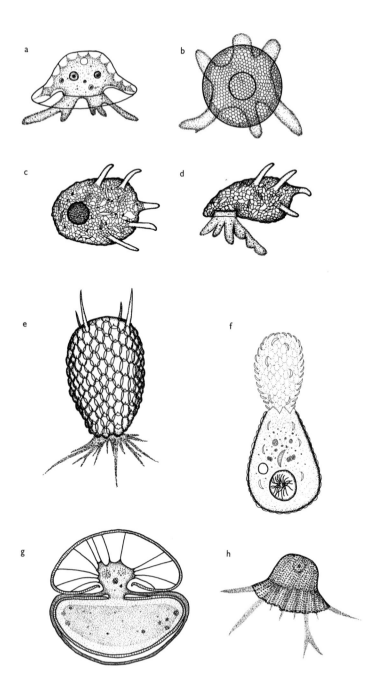

a

b

c

d

e

f

g

h

Fig. 7.11 *Euglypha rotunda.* a. Scanning electron micrograph of the shell, which is formed of imbricated siliceous scales. b. Electron micrograph of an oblique longitudinal section showing the scaly shell enclosing the cell contents; the nucleus and dense cytoplasm are seen above, while vacuoles, loose scales and a few mitochondria may be seen below (p. 170). Micrographs by R. H. Hedley [101a].

Fig. 7.10 Testate amoebae. *Arcella vulgaris* (50 μm) showing features of the shell and pseudopodia (*a*) from the side and (*b*) from above; at *g*, the formation of a new shell before division of *Arcella* is shown—much of the cytoplasm emerges from the upper parent shell and a new shell is secreted beneath the surface of the protruded cytoplasm; later the cytoplasm divides and the two shelled progeny separate (information from Netzel).[185] *Centropyxis aculeata* (100 μm) showing the shell from below (*c*) and the shell with pseudopodia from the side (*d*). *Euglypha alveolata* (150 μm) (*e*) with siliceous scales, spines and pseudopodia extending from the aperture; *f* shows the formation of a new shell of *Euglypha* which takes place prior to binary fission by the fitting together of the preformed scales in an imbricated array at the surface of a mass of extruded cytoplasm. *Cochliopodium bilimbosum* (40 μm) (*h*) has a flexible test coated with scales reminiscent of those on some flagellates.

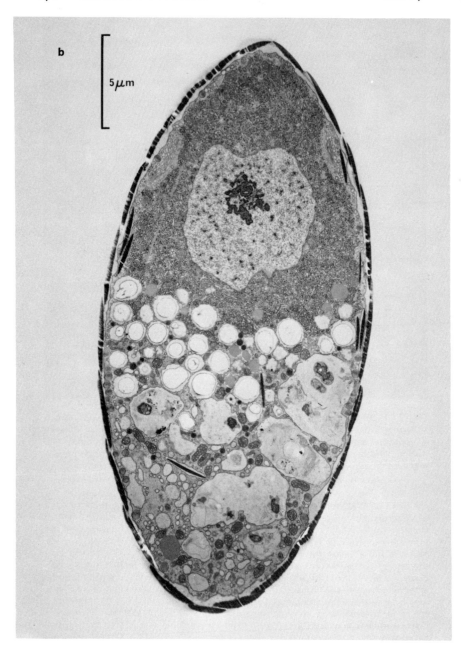

Fig. 7.11b   (see legend p. 169).

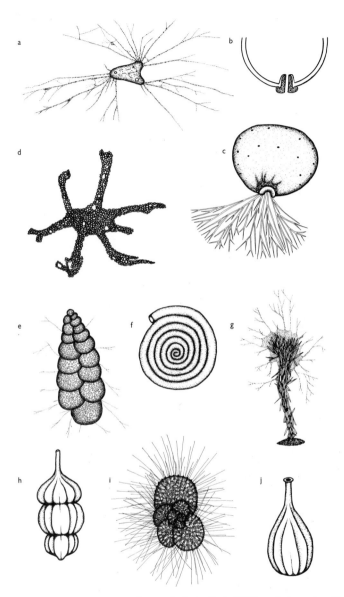

Fig. 7.12 Foraminiferans; a, *Allogromia laticollaris* (600 μm); b and c, *Gromia oviformis* (2 mm), with details of the shell aperture shown in b; d, *Astrorhiza limicola* (15 mm); e, *Textularia agglutinans* (1 mm); f, *Cornuspira involvens* (1 mm); g, *Haliphysema tumanowiczii* (2 mm); h, *Nodosaria* sp. (800 μm); i, *Globigerina bulloides* (600 μm) complete with spicules (cf. Fig. 7.17); j, *Lagena striata* (500 μm).

minute surface scales of the flexible test of *Cochliopodium* (Fig. 7.10h) are reminiscent of those found on many phytoflagellates (p. 116). Pseudopodia are used for the capture of food, principally diatoms and other algae, and for locomotion. During binary fission one daughter retains the original shell, while the other constructs a new one, either from the collection of preformed scales in *Euglypha* or from sand grains collected in the cytoplasm or around the aperture in *Difflugia*; the new shell is formed mouth to mouth against the existing one (Fig. 7.10f, g).

Fig. 7.13 Three specimens of *Julienella foetida* (with cm scale); this foraminiferan is a dominant member of an ecological community on silty sand off the coast of Ghana. Specimens collected by R. Bassindale.

Foraminiferans[101, 151, 180]

The best-known of the granulo-reticulose amoebae are members of the order Foraminiferida, to which about one half of all named Protozoa belong. They build shells or tests composed of one or many chambers from organic secretions, often with added arenaceous or calcareous materials. The shells of many foraminiferans provide good fossils which are abundant in marine and brackish water deposits from the palaeozoic onwards; their small size and considerable abundance make them valuable to strati-

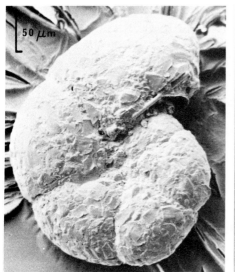

Fig. 7.14 Scanning electron micrographs of the foraminiferans *Cribrostomoides jeffreysii* (a) with agglutinated fragments of sand incorporated in the shell, and *Oolina melo* (b) with sculptured proteinaceous shell. Micrographs by J. W. Murray.[180]

graphers, and they have proved of particular importance to the oil industry in the identification of strata in rock drillings. A large proportion of named foraminiferans are fossil forms.

Single-chambered membranous tests are formed by such foraminiferans as *Allogromia* (Fig. 7.12a), which may be found in the holdfasts of laminarian algae. The principal material of the test is 'tectin' (glycoprotein of the acid mucopolysaccharide type) together with some other material, including protein, lipid and sometimes an impregnation of ferric iron. The internal cytoplasm emerges at one or more specialized apertures to form the pseudopodial reticulum and often also a layer of cytoplasm over the outside surface of the test. *Gromia*, whose pseudopodia are described as filose, has a very similar but rather thicker tectinous shell with a single specialized aperture[102] (Fig. 7.12b, c), but is frequently classed in a group separate from foraminiferans.

Shells in which arenaceous material is incorporated may have one or several chambers, and may have regular or irregular shapes—in some cases they may be several cm across (Fig. 7.13). It is thought that in at least some cases the shell is built from a tectinous, organic layer upon which an arenaceous layer is superimposed, but in other forms these components seem to be completely mixed. Arenaceous shells may be hard and brittle or soft and flexible; in both cases the cement between the foreign particles is basically acid mucopolysaccharide, but this is mineralized with calcium and iron salts in the rigid shells. The particles that are cemented together may be an unselected collection of mineral grains covering a considerable size range, as in *Astrorhiza* (Fig. 7.12d) or *Cribrostomoides* (Fig. 7.14a), or may be a very selective collection of particles, e.g. echinoderm plates in *Technitella thompsoni*, or sponge spicules in *Technitella legumen*. Sponge spicules form the main part of the test in *Haliphysema* (Fig. 7.12g), which is commonly found among tufts of algae and Bryozoa on British coasts.

Foraminifera with calcareous shells mostly have many chambers whose walls are formed from an organic layer reinforced with calcite. The shells of these forms are composed either of minute crystals which give the shells

Fig. 7.15 Photomicrograph of a miliolid foraminiferan showing the pseudopodia emerging from the shell aperture.

a shiny appearance like porcelain, or of minute granules or radially arranged crystals, which form hyaline shells. In a few families the shells are made of arragonite or of silica. The porcellanous shells are normally without perforations and all pseudopods emerge from the single aperture, e.g. miliolids (Fig. 7.15), while the hyaline shells are normally perforated and small pore pseudopodia emerge through the perforations as well as the larger pseudopodia from the aperture (Fig. 7.16). Some examples of these foraminiferans are shown in Fig. 7.12f, h, i and j; they seldom exceed one or two mm in diameter.

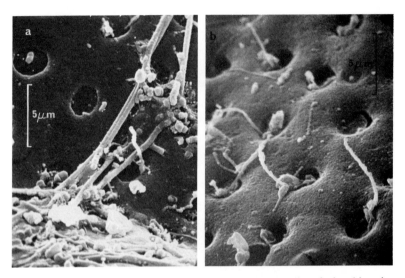

Fig. 7.16 Scanning electron micrographs of pseudopodia of *Amphistegina madagascariensis*. In (*a*) normal pseudopods are seen passing over the perforated shell surface, which also bears fragments of debris; in *b* small pseudopods are seen to emerge from pore plates at the mouths of shell perforations. Micrographs by H. Jørgen Hansen.[128]

The majority of foraminiferans live on the sea floor, particularly in shallower waters where they are common on weeds, rocks and softer substrates. There are also abundant and important planktonic foraminiferans, such as *Globigerina* (Figs. 7.12i, 7.17), which has radiating calcareous spicules to aid flotation. They feed on almost any form of food that may come in contact with their adhesive pseudopodia—from unicellular algae, yeasts and bacteria to copepods and other crustaceans. While bacteria and small food particles are carried with the cytoplasm of the pseudopods, larger bodies of food are surrounded by numbers of

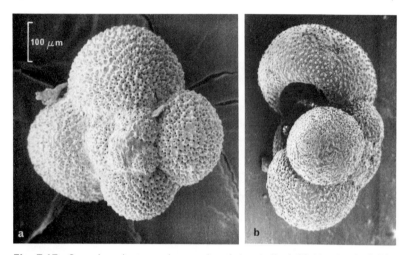

Fig. 7.17 Scanning electron micrographs of the shell of *Globigerina bulloides* from above (*a*) and from the side (*b*) showing the aperture. Micrographs by J. W. Murray.[180]

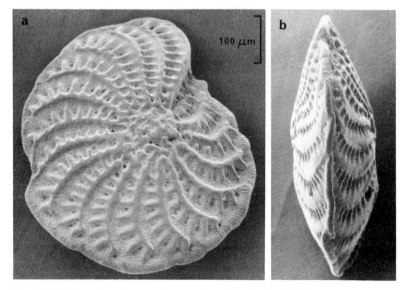

Fig. 7.18 Scanning electron micrographs of the shell of *Elphidium crispum* seen in side view (*a*) and from the edge (*b*). Micrographs by J. W. Murray.[180]

pseudopods and may be carried within the shell of the foraminiferan, where, according to one report, the food particles are digested extracellularly in lacunae.[156] Certain foraminiferans regularly contain symbiotic algae. Some information on the life cycles of foraminiferans was given in Chapter 4 (p. 84); relatively few species have been studied, and these are quite diverse, although all of them seem to have an intermediary meiosis that is unique among animals. Dimorphism of the shell occurs in some species with multichambered shells; the first chamber (proloculus) is smaller in shells of one generation than in the other, and it has been established that in *Elphidium* (Fig. 7.18) the zygote forms the smaller proloculus found in the agamont generation, while in *Spirullina* the shells with the smaller proloculus belong to the gamont generation.

The Actinopodea[261]

The Heliozoa, Radiolaria and related forms comprise the Actinopodea, a distinct section of the amoeboid Protozoa consisting typically of spherical floating forms with radiating pseudopods.

The most familiar **heliozoans** are *Actinophrys* (Fig. 7.19a) and *Actinosphaerium*, which are common in freshwater; the former has a single central nucleus and the latter is multinucleate. The nuclei lie in the endoplasmic region, which is granular and contains small vacuoles, while the ectoplasmic region is a frothy layer of large vacuoles, some of which are contractile. The cytoplasm also contains food vacuoles within which flagellates, algae, ciliates and even rotifers may be digested. The structure of the characteristic axopodia and their role in movement has been described on p. 160. Binary fission is the normal mode of reproduction, but a sexual process of autogamy also occurs (p. 83).

In the **centrohelid heliozoans**, such as *Acanthocystis* (Fig. 7.19b) and *Raphidiophrys*, the axial fibrils of the axopodia arise from a central granule or centroplast (Fig. 7.7), rather than from a nuclear membrane, so that the nucleus is eccentric; the centroplast has at least some of the properties of a centriole, including participation in the formation of the mitotic spindle. Many of these forms have siliceous skeletal elements in the form of radiating spines or tangential peripheral scales, and the axopodia of *Raphidiophrys* have been found to carry bodies similar to the haptocysts of suctorian ciliates (p. 64). Centrohelids are found in freshwater and sometimes in salt water, but have not been extensively studied. Even less well known are those species with a more complete silica skeleton formed into a spherical reticulum, e.g. *Clathrulina*,[16] which is a stalked species found attached to plants in freshwater (Fig. 7.19c), and which has a biflagellate migratory phase in the life cycle. Small protozoans with radiating pseudopods, described as filopodia or axopodia, and also bearing

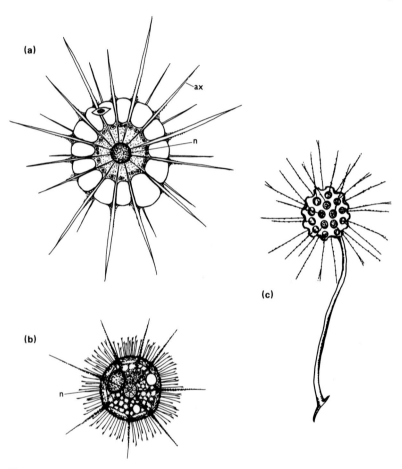

Fig. 7.19 Heliozoans; *a, Actinophrys sol* (40 μm) with nucleus (n), axopodia (ax) and vacuolated surface; *b, Acanthocystis chaetophora* (50 μm) with nucleus (n) at one side of the centroplast (Fig. 7.7) and siliceous bifid spines as well as axopodia; *c, Clathrulina elegans* (diameter 75 μm).

Fig. 7.20 Radiolarians and Acantharia; *a, Aulacantha scolymantha* (1 mm) with a skeleton of separate radiating rods, a prominent central capsule containing the nucleus and surrounded by a region of more granular cytoplasm with numerous oil droplets enclosed within the vacuolated calymma zone; *b, Acanthometron elasticum* (300 μm) with regularly arranged radial spines, and a central capsule containing zooxanthellae; *c,* skeleton of *Hexacontium asteracanthion* (120 μm); *d,* skeleton of *Pipetta fusus; e,* skeleton of *Cycladophora pantheon; f,* small colony of *Collozoum inerme* (up to 10 mm or more long) with numerous central capsules in a common vacuolated calymma zone.

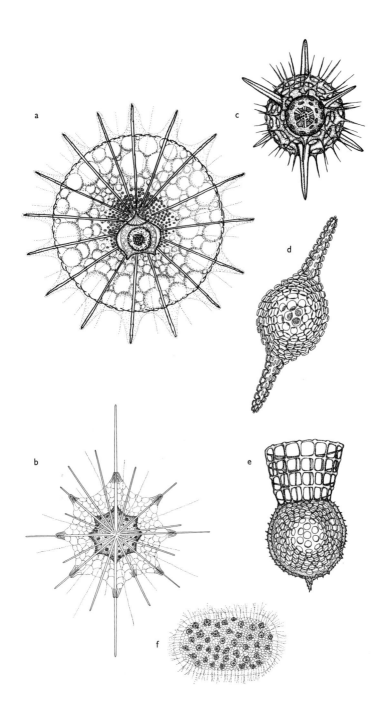

a

c

d

b

e

f

flagella, are sometimes referred to as helioflagellates; these organisms are not well known, but include *Actinomonas* with one flagellum (Fig. 5.10g, p. 121) and *Dimorpha* with two flagella and axopodia that radiate from a 'centriole'.

Most **radiolarians** are pelagic marine forms, while the remainder are benthic; because of the oceanic habitat of the Radiolaria their biology

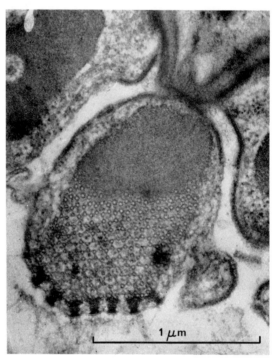

Fig. 7.21 Electron micrograph of a near-transverse section at the base of an axopodium of the radiolarian *Sticholonche zanclea*; the hexagonally arranged microtubules of the axopodial axis originate from a dense region which articulates at a specialized socket at the body surface. Micrograph by A. Hollande, J. and M. Cachon and J. Valentin.[106]

is less well known than that of Heliozoa.[67] The cells of larger radiolarians are several mm across and such colonial species as *Collozoum* may reach a size of 10 or 20 cm. Their skeleton is internal and siliceous. Radiolarian skeletons predominate in the ooze of certain areas of the ocean floor and are also found frequently as fossils in certain rocks. Within the complex body the cytoplasm is separated into two regions by a membranous central capsule (Fig. 7.20a), which may be single or double, and which is perforated by one, a few or many pores. The cytoplasm within the capsule is granular and somewhat vacuolated; it contains one or more nuclei and food storage materials, principally oil and fat droplets. Outside the central capsule is a thin layer of compact cytoplasm surrounded by an extensive, highly-vacuolated region, the calymma. Within the calymma are food

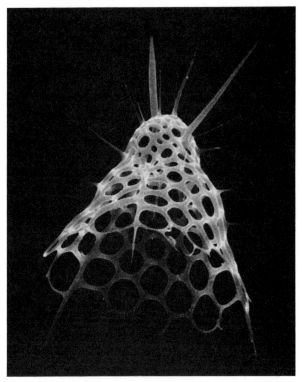

Fig. 7.22 Scanning electron micrograph of the silica skeleton of the radiolarian *Pterocanium trilobium* (about 85 μm long). Micrograph by J. and M. Cachon.[29a]

vacuoles containing planktonic animals and plants that are caught by the pseudopodia, and also in many species of radiolarians symbiotic algae (zooxanthellae) live in the calymma region and apparently enable these radiolarians to survive long periods in light without catching prey. The vacuolated layer is believed to provide a variable buoyancy which can be controlled to achieve a certain amount of vertical movement or depth regulation. Pseudopodia, in the form of fine filopodia or axopodia (Figs. 7.5d and 7.21), extend outwards from the calymma and presumably also aid in flotation as well as food capture. The skeleton is composed primarily of spicules, frequently radially or tangentially disposed, but not penetrating the central capsule. Elaborate spherical or less regular skeletons may be formed within the cytoplasm by the fusion of spicules. These skeletons are among the most exquisite of natural structures and are often exceedingly complex; some of them consist of several concentric spheres, representing successive stages of growth of the organism—the central capsule may grow to enclose one or more of these spheres, and the central sphere may become intranuclear. Some examples of these skeletons are shown in Figs. 7.20 and 7.22.

A similar organization of the body is found in the **Acantharia**, whose radiating spicules are said to be composed of strontium sulphate, although a recent report states that the material is really a mixture of silicates of calcium and magnesium. In *Acanthometron* (Fig. 7.20b) the 20 precisely arranged radial spicules meet at the centre of the organism after penetrating the central capsule. The zooxanthellae in these forms are found within the central capsule, but in other respects these amoebae appear like radiolarians.

Reproduction in both Radiolaria and Acantharia is imperfectly known; it is said to involve multiple division to produce biflagellate zoospores, which may represent gametes.

8

The Ciliate Protozoa

Ciliate Protozoa are characterized by the possession of two types of nucleus, the sexual process of conjugation (p. 87), a basically equatorial division plane in binary fission (p. 98), and a form of pellicular organization in which cilia and their intracellular attachments (the infraciliature) are dominant features. There has been extensive evolutionary exploitation of the ciliate pattern of organization, and this is most clearly reflected in changes in the ciliature and infraciliature. There is also evidence that the characteristic nuclear dualism evolved within the group after the appearance of the main features of pellicular structure.

THE NUCLEI OF CILIATES.[209]

Most ciliate Protozoa possess nuclei of two types, named macronuclei and micronuclei. Viable strains of ciliates without micronuclei are known, but with the exception of *Stephanopogon* (see below) all ciliates are thought to possess at least one macronucleus and in some ciliates many macronuclei are present; the number of micronuclei is often greater than the number of macronuclei. The micronuclei are small and compact and are normally diploid with large amounts of histone protein but little or no RNA (no nucleoli). At binary fission micronuclei divide mitotically within persistent nuclear membranes, and in the sexual processes of conjugation and autogamy (p. 87) a micronucleus undergoes meiosis before giving rise to the gametic pronuclei. Following sexual processes it is normal for the macronucleus to degenerate and for a new macronucleus to be formed from division products of the micronuclear syncaryon (p. 88). Macronuclei are larger and are normally highly polyploid, estimates of ploidy levels of 16 n to 13 000 n having been recorded from polyploid macro-nuclei of various ciliates. During the formation of a macronucleus from a micronuclear syncaryon, rapid replication of chromosomes occurs and the macronucleus becomes polyploid, with in some cases a stage when poly-tene chromosomes are visible. In the fully formed macronucleus the chromatin is confined to many small dense bodies containing numerous 10 nm diameter microfibrils twisted together as in the chromosomes of other organisms; the numerous, larger, less-dense nucleoli contain much RNA and ribosome-like bodies. It has been shown that protein synthesis in the cytoplasm depends on RNA of macronuclear origin. Formation of DNA as well as RNA occurs in most macronuclei, and in some examples endomitotic replication of macronuclear chromosomes has been observed in which distinct threads appear, split longitudinally, and complete separa-tion shortly before the macronucleus divides. In an excellent recent review concerning macronuclei I.B. Raikov[209] quotes evidence that during macro-nuclear division many complete genomes become segregated into each daughter nucleus; this segregation of complete genomes is thought to be achieved by the joining together of members of each genome to form linear composite chromosomes which replicate as a unit and segregate as a unit. During division of the macronucleus, which takes place within an intact nuclear membrane, the nucleus elongates by the extension of a 'pushing spindle', which is a bundle of microtubular fibrils that correspond with the continuous fibres of a normal mitotic spindle.[262] Following the sexual processes of conjugation or autogamy the parental macronucleus degener-ates, often with fragmentation, and is replaced by a newly formed macro-nucleus; normally this is provided by division of the micronuclear syn-caryon, but, in the absence of such a macronuclear anlage, one or more fragments of the degenerating parental macronucleus may regenerate to form a complete new macronucleus.

It is valuable to compare this typical pattern of behaviour of the dual nuclear structures in ciliates with the patterns shown by the nuclei of *Stephanopogon* and by the ciliates with diploid macronuclei (Fig. 8.1). *Stephanopogon* is a member of the ciliate group that is regarded as ancestral to all other ciliates (the rhabdophorine gymnostomes, see p. 194). In *S. colpoda* young individuals have two identical nuclei, each of which has a large central nucleolus and peripheral chromatin and appears rather like the nucleus in some of the smaller amoebae. During growth both nuclei undergo several mitoses that involve constriction of the nucleolus without its dissolution, followed by anaphase migration of chromosomes. Large individuals have up to sixteen nuclei and following encystment the cyto-plasm divides into uninucleate masses, within each of which mitosis takes place so that at excystment a number of binucleate ciliates emerge. Sexual processes have not been observed.

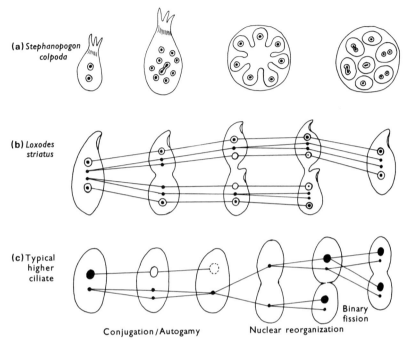

Fig. 8.1 Comparison of the behaviour of nuclei during the life cycles of three ciliates. *Stephanopogon colpoda* (a) has a single type of nucleus (see text). In *Loxodes striatus* (b) both the macronuclei and micronuclei are diploid, but macronuclei never divide, so that new macronuclei are always formed from micronuclei and additional micronuclear divisions are necessary to maintain the full number of nuclei. In a typical higher ciliate (c) the macronucleus disappears during conjugation, during which the micronucleus undergoes meiosis and fertilization, and then a new macronucleus is formed from the micronuclear syncaryon; during binary fission both macronuclei and micronuclei divide. Information for (a) and (b) from I. B. Raikov.[209]

In three families of lower holotrich ciliates, and a number of genera of uncertain affinity, the macronuclei are diploid, and are unlike the macronuclei of typical ciliates in both structure and behaviour. The micronuclei of these ciliates are typical and lack RNA, while the rather small macronuclei contain the same amount of DNA as is present in micronuclei during their G_2 period, and generally have a large central nucleolus and peripheral granular chromatin (as in *Stephanopogon* and some amoebae). Synthesis of RNA appears to be confined to these diploid macronuclei, but synthesis of DNA only takes place in the micronuclei. The macronuclei, therefore, never divide; they are segregated into the daughter cells at cell division, and their number is maintained by transformation of micronuclei directly to macronuclei by development of nucleoli and condensation of chromatin. During the interphase of each cell cycle the micronuclei undergo several mitoses to maintain the number of macronuclei and the number of micronuclei; both mitosis of micronuclei and transformation of micronuclei to macronuclei may occur at any part of the cell cycle. There may be several macronuclei and several micronuclei, and the number of macronuclei may be greater than the number of micronuclei so that many micronuclear mitoses are necessary to maintain the complete nuclear complement. A species with two nuclei of each type is *Loxodes striatus* (Fig. 8.1b), while *Trachelocerca coluber* has 4 macronuclei and 2 micronuclei, and up to thirty-five macronuclei have been found in *Remanella multinucleata*.

In discussing the origin and evolution of nuclear dualism in ciliates, Raikov points out that organisms exist (*Stephanopogon* spp) which have developed the characteristic somatic ciliature of ciliated Protozoa, yet have primitive nuclei of a single type. This suggests that the nuclear dualism seen in other ciliates arose within the group. There is however the tendency towards multinuclearity that also characterizes evolution in most other protozoan groups in association with increase in body size and increased synthetic activity. This could be followed by a primary differentiation into generative nuclei (micronuclei) and somatic nuclei (macronuclei), with initially a pattern in which the generative nuclei retain the ability to divide to produce micronuclei and macronuclei, while the somatic nuclei lose the power to divide. This incidentally is the stage of nuclear dualism reached by some foraminiferans, and may be valuable because it permits the nuclear material to become more active in the synthesis of RNA without the need to conserve the ability to replicate DNA. The formation of more conventional macronuclei involved the replication of genomes in somatic nuclei leading to the polyploid state, which required the reacquisition (by retention from the micronuclear progenitor) of the ability to replicate DNA within the macronucleus and which is associated with the ability to divide the nucleus by segregation of genomes. Nuclear dualism with polyploidy as seen in ciliates provides a more advantageous mechanism than polyploidy without nuclear dualism, as seen in Radiolaria, because of the presence of the separate genetic store of the micronuclei.

ORGANIZATION OF THE CILIATE PELLICLE[89, 198, 200]

Pellicle structure in *Tetrahymena*.

In many ciliates the cilia are only present as feeding organelles or on part of the body surface, but this is believed to result from reduction and specialization of the ciliature. The pellicular structure of *Tetrahymena* provides a good illustration of the arrangement of cortical organelles in a ciliate. *Tetrahymena* is a most important ciliate with unspecialized features. It occupies a central position in ciliate classification, and has probably been the subject of more scientific papers than any other ciliate, primarily because it has been maintained in axenic culture for several decades and has been found to be excellent material for biochemical studies.[41] The body ciliature of *Tetrahymena* shows features found in many of the more primitive ciliates and the basic components of the mouth ciliature of more complex ciliates are represented in the buccal cavity of *Tetrahymena*.

The body cilia of *Tetrahymena* occur in longitudinal rows called kineties.[2, 3] Each ciliary base is associated with several intracellular fibrils and a pair of membranous vesicles to form the basic unit of pellicular organization which is called the kinetid and occupies a 'ciliary territory' (Fig. 8.2). A surface membrane covers the whole organism, including the cilia (p. 28); beneath this membrane in most places are two further membranes which can be seen to form the outer and inner walls of flattened vesicles or alveoli extending over the body surface, and arranged in pairs around the base of each cilium, one on the left side and one on the right. The outer vesicle membrane lies close to the cell membrane, and the space within the vesicle is variable, perhaps as a result of different techniques of fixation for electron microscopy. The membranes between adjacent vesicles are seldom complete, and the alveolar cavities of a whole kinety may be continuous. The innermost membrane is underlain by a dense layer of cytoplasm called the epiplasm.

The basal body (kinetosome) of the cilium is closely surrounded by the membrane vesicles at its outer end near the transition from basal body to cilium, and at its inner end is associated with three ectoplasmic fibril systems. The largest fibril is that which contributes to the kinetodesmos, a fibre which runs along the (animal's) right side of the row of kinetosomes of a kinety. A kinetodesmal fibril arises at the right of the anterior aspect of the kinetosome, being connected to one or more of triplets 5, 6, 7, and 8 by amorphous material. From this point the fibril passes forwards towards the right and outwards towards the innermost membrane of the pellicle, where it runs alongside the kinetodesmal fibril of the next anterior kinetid before terminating. The kinetodesmos is not normally a continuous fibre as it may appear in the light microscope, but is merely a series of overlapping fibril units. Each fibril is striated with a periodicity of about 26 nm (repeat periods between 15 and 50 nm have been reported from different ciliates), and is made up of a close bundle of longitudinal filaments; it is widest at the kinetosome and tapers gradually towards its anterior extremity.

A set of 7 to 12 longitudinal microtubular fibrils is found between the

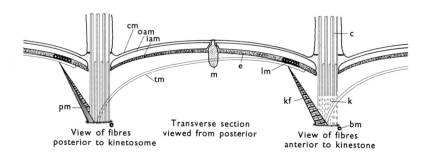

View of fibres posterior to kinetosome

Transverse section viewed from posterior

View of fibres anterior to kinestone

Surface view of fibre systems, pellicular alveoli and mucocysts shown dotted

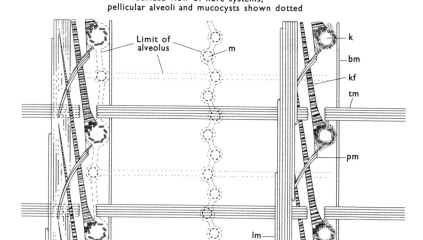

Fig. 8.2 Pellicle structure in *Tetrahymena*, seen in transverse section (above) and in surface view (below). *bm*, Basal microtubules: *c*, cilium: *cm*, cell membrane: *e*, epiplasm: *k*, kinetosome: *kf*, kinetodesmal fibre: *iam* and *oam*, inner and outer alveolar membranes: *lm*, longitudinal microtubules: *m*, mucocyst: *pm*, post-ciliary microtubules: *tm*, transverse microtubules. Cilia of adjacent rows are not necessarily opposite, they are often placed alternately.

innermost pellicular membrane and the epiplasm a little to the right of the kinetodesmos. These fibrils also do not run the length of the organism, but form an overlapping series of shorter fibrils. They do not make any direct connection with the kinetosomes, although they are associated with two other sets of microtubular fibrils which do arise near ciliary basal bodies.

At the anterior of the kinetosome on the left a group of 5 or 6 transverse microtubules is seen in association with amorphous material around triplets 4 and 5 of the kinetosome; these microtubules run outwards towards the pellicle and transversely towards the left, so that they pass immediately below the epiplasm and terminate close to the longitudinal microtubular fibrils near the next kinetodesmos on the left. Another group of 4 to 8 postciliary microtubules originates at the posterior side of the kinetosome on the right close to triplet 9 and runs towards the pellicle and backwards to the right, where they also terminate close to the longitudinal microtubules. One or several basal microtubular fibrils have also been found running longitudinally beneath and to the left of the kinetosomes of a kinety and making contact with each basal body.

The association of the terminations of the transverse, postciliary and kinetodesmal fibrils with the epiplasm and longitudinal microtubules of the pellicle, and also with the dense material around the inner end of the kinetosome, suggests an anchorage function for these fibres. These fibril systems, together with the attachment of the distal end of the basal body to the pellicular membranes at the cell surface, can form a firm tetrahedral anchorage for the ciliary base. The basal microtubules are not as straight as those of the other systems, and may have some other function. The regularly arranged ciliary territories are linked in rows by the kinetodesmal fibrils and by their relation to the longitudinal bands of microtubules, so that these fibres may also have an important morphogenetic function.

Immediately anterior to most kinetosomes are invaginations of the surface membrane called parasomal sacs, which extend inwards for one third to one half the length of the basal body. Beneath the spaces between pellicular alveoli which lie along primary meridians in line with the kinetosomes and along the secondary meridians between the kineties are bodies called mucocysts, which make contact with the surface membrane when they are mature, and are believed to be the source of a mucoid secretion onto the body surface.[223a]

Pellicle structure in other ciliates

Patterns comparable with this are found in other ciliate Protozoa, Fig. 8.3. In many other holotrich ciliates the kinetodesmal fibres are larger and longer; e.g. in *Paramecium* (Fig. 8.3b) they extend anteriorly past 4 or 5 cilia before terminating,[4, 199] and they are even longer in some astome and apostome ciliates, so that the kinetodesmos in these forms is a much larger bundle of fibrils.[207] In large species of *Paramecium* and many other ciliates two mature kinetosomes may occur in each kinetid; the kineto-desmal fibrils and postciliary microtubules are then associated with the posterior kinetosome which may or may not carry a cilium and the trans-verse microtubules with the anterior kinetosome which always carries a cilium. Groups of microtubules in the position of the basal microtubules of *Tetrahymena* are more important in some cases, for example in cyrtophorine gymnostomes[160] (Fig. 8.3c) and thigmotrichs.[134] Three or more rows, each with 5 or 6 microtubules, may underlie the kinetosomes of a kinety; such

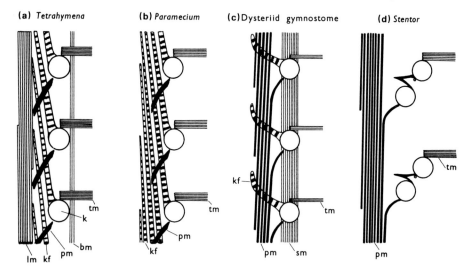

Fig. 8.3 Comparison of patterns of ciliary root fibres in four ciliates. *sm*, Sub-kinetal microtubules; other abbreviations as in Fig. 8.2. (Surface views.)

subkinetal microtubules originate from kinetosomes and extend forward past several kinetids. In these cases the postciliary and subkinetal micro-tubules are the best developed longitudinal elements of the infraciliature. In such heterotrich ciliates as *Blepharisma*,[131] *Stentor*[11] and *Spirostomum*[73] microtubular fibrils are also the dominant components of the infraciliature. In these species a row of 9 or more microtubules arises near triplet 9 of the posterior kinetosome (in the position of the postciliary microtubules) and runs backwards on the right to join with several other rows of microtubules from more anterior kinetids to form a stack of rows of microtubules (the km fibres of some authors) (Figs. 8.3d, 8.4). This large bundle of micro-tubules lies along the right side of the row of kinetosomes in the position

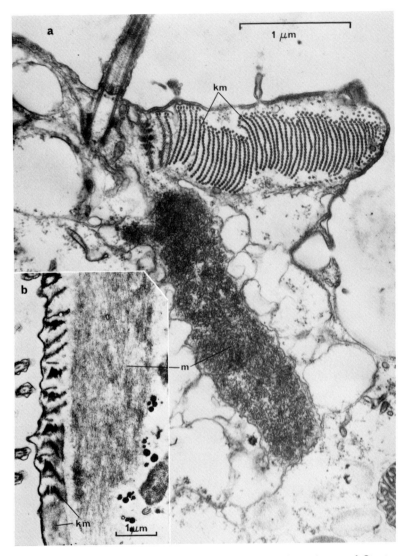

Fig. 8.4 Electron micrographs of sections of contracted specimens of *Stentor coeruleus* to show the structure of pellicular fibre systems. a, Transverse section passing through a ciliary base, a stack of rows of postciliary microtubules (km) and a myoneme (m); in the longitudinal section (b) portions of the km and m fibres are seen. Micrographs by L. H. Bannister and E. C. Tatchell.[11]

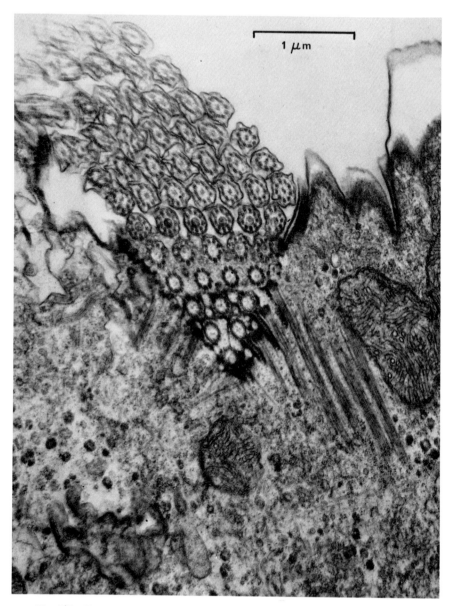

Fig. 8.5 Electron micrograph of a section passing obliquely through the base of a cirrus of *Euplotes eurystomus* showing the arrangement of the kinetosomes and the microtubular root fibrils. Micrograph by R. Gliddon.

of the striated kinetodesmos of other ciliates—from light microscopy it has been called a kinetodesmos, but is seen from electron micrographs to have a different origin and structure. In such contractile ciliates as *Stentor* and perhaps *Spirostomum* it is suspected that these microtubular bundles are concerned with the extension of the animal following contraction.[11, 116] Contraction in these two ciliates is believed to be caused by the M fibres which are large bundles of microfilaments situated deeper in the body than the microtubular fibrils (Fig. 8.4).[73, 155] These bundles may be an elaboration of the infraciliary lattice of microfibrillar tracts which has been found in *Paramecium* at the boundary between the ectoplasm and endoplasm, with branches in the ectoplasmic region. The M fibres (myonemes) of *Stentor* are connected to each other by lateral branches, and fine filamentous material connects them with the kinetosomes and with the pellicle. Excitation of the contraction of these myonemes is believed to result from a raised concentration of free calcium ions in the cytoplasm.[73, 236] Transverse microtubular fibrils are present in these heterotrichs, but it is not known where they terminate.

Body ciliature may be reduced to specialized cilia, such as the dorsal bristle cilia and ventral cirri of hypotrich ciliates like *Euplotes*.[87] In this case the cirrus kinetosomes carry large bundles of microtubular fibrils (Fig. 8.5); these run in the ectoplasm but possible homologies with transverse, postciliary or subkinetal fibril systems of *Tetrahymena* cilia are not clear at this stage. In other species the body cilia may be missing, at least for most of the life of the ciliate, as in peritrichs like *Vorticella*, where, however, the kinetosomes of a girdle of cilia remain beneath the pellicle and develop ciliary shafts under appropriate conditions (p. 209). In suctorians the sessile adult lacks cilia, but gives rise to a motile larva whose cilia develop from kinetosomes that had lain dormant beneath the pellicle of the adult ciliate.

In *Paramecium* and some other ciliates the trichocysts occur in similar positions to the mucocysts of *Tetrahymena*; mature undischarged trichocysts of *Paramecium* show a paracrystalline structure of protein material which suddenly elongates on discharge with extension of the paracrystalline lattice by a factor of about 8.[99, 249]

MOUTH CILIATURE AND THE
EVOLUTION OF CILIATES[38, 40, 200]

While the fibrils associated with somatic kinetosomes generally remain in the ectoplasmic region of the ciliate, the kinetosomes of mouth cilia give rise to fibres which may run deep into the endoplasm of the cell. In many ciliates plates of dense material at the inner end of kinetosomes in the oral area act as the origins of bundles of hexagonally packed microtubules called nemadesma, possibly homologous with subkinetal microtubules, which extend for a variable distance into the endoplasm. Other rows of microtubules, called cytopharyngeal ribbons, and interpreted as belonging to either the post-ciliary or transverse categories of microtubules, are found beneath the surface membranes around the cytostome, and may subsequently extend into the endoplasm of the cytopharynx region. A filamentous reticulum which may represent a specialization of the infraciliary lattice of the somatic regions often occurs in close association with kinetosomes and other organelles of the mouth region; it forms a close meshwork with dense nodes that is best seen in peritrichs, where it is believed to support the walls of the buccal cavity. At the cytostome the body surface is formed by a single unit membrane, the complex pellicle structure being modified around the mouth and interrupted at the cytostome.

The ciliature of the mouth region is more diverse than the somatic ciliature, but shows more consistency of pattern within the various ciliate groups, and so is more important than the body ciliature as a taxonomic feature. For this reason a description of the types of oral ciliature will be coupled with a brief reference to the orders of the subphylum Ciliophora*.

In the simplest ciliates, members of the **Gymnostomatida**, the mouth shows no compound ciliature, although some perioral cilia may assist in feeding and may give rise to nemadesmata and other fibres. The cytostome is at the body surface or is only slightly recessed and often leads into a well developed cytopharynx whose walls are supported by nemadesmal rods. In the rhabdophorine gymnostomes there may be several concentric cylinders of such rods flexibly arranged to permit the ingestion of large food masses, which may include other Protozoa, algae or organic detritus. These forms take in this food through a slit mouth on the proboscis region in *Amphileptus* or *Loxophyllum* (Fig. 8.6a, b), a circular mouth at the base of the proboscis in *Dileptus* (Fig. 8.6c), or in more specialized predators like *Didinium* (Fig. 8.7) there may be a highly developed proboscis incorporating nemadesmal and other components as well as specialized trichocysts.[288] In species of *Coleps* an array of pellicular plates gives the barrel-shaped body a characteristic sculptured appearance (Fig. 8.8a). In the herbivorous cyrtophorine gymnostomes (also called the Hypostomata) the mouth is moved ventrally rather than laterally, and within the cytostome is a cytopharyngeal basket formed from a close cylindrical array of nemadesmata (see p. 67), through the centre of which diatoms or filamentous algae are taken in, e.g. *Nassula, Chilodonella* (Fig. 8.8b).

Ciliates of the **Trichostomatida** show better developed feeding ciliature because the body surface around the cytostome forms a depression called the vestibulum into which rows of kinety cilia extend. The cilia in the vestibulum may be longer and closer together than those of the ordinary body surface, but are never formed into compound structures.

* A comprehensive revision of the classification of ciliates has been proposed.[294]

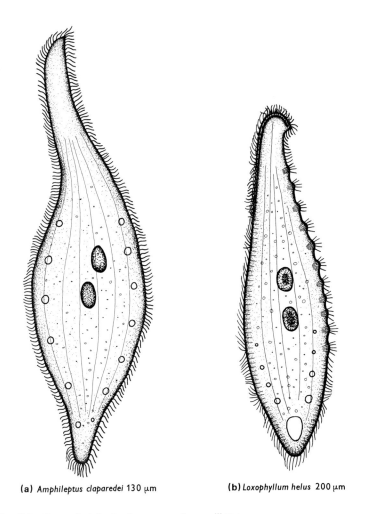

(a) *Amphileptus claparedei* 130 μm   (b) *Loxophyllum helus* 200 μm

Fig. 8.6 Some rhabdophorine gymnostome ciliates.

In *Balantidium* (Fig. 8.9a), for example, there are many close packed kineties at the left side of the vestibulum, and usually shorter ones at the right side, the dorsal area being naked. Nemadesmata arise from kinetosomes on both sides of the vestibulum and may form an extensive cylinder around the cytopharynx. Rows of microtubular fibrils arising from the terminal kinetosomes at the left side are found as cytopharyngeal ribbons beneath the membrane of the dorsal area and extending to the cytopharynx.

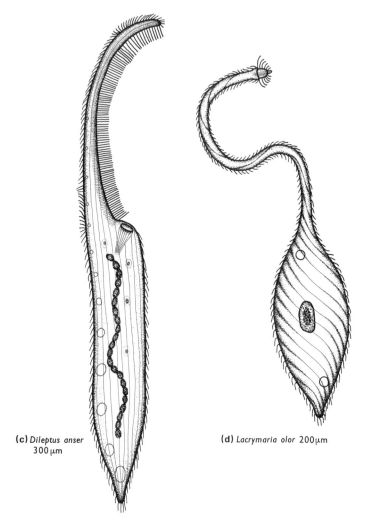

(c) *Dileptus anser* 300 μm

(d) *Lacrymaria olor* 200 μm

These ciliates commonly feed on small particles and bacteria. In addition to such endocommensals as *Balantidium* and *Isotricha*, the common soil species of *Colpoda* (Fig. 8.9b) are usually placed in this order.

Another group which is thought to be derived from the gymnostomes is the order **Chonotrichida**, members of which are sedentary ectocommensals on Crustacea. The vase-shaped body of these forms lacks ordinary body cilia, but within the funnel-shaped apical vestibulum rows of cilia

Fig. 8.6 (cont.)

run down towards the cytostome. Lateral budding gives rise to a ciliated migratory larva. Most species of chonotrichs are marine, but the best known form is *Spirochona* (Fig. 8.10), which is common on the gills of freshwater species of *Gammarus*. The macronuclei of chonotrichs show a subdivision into a DNA-rich region and a region lacking in DNA; this type of structure is only found elsewhere in certain cyrtophorine gymnostomes.

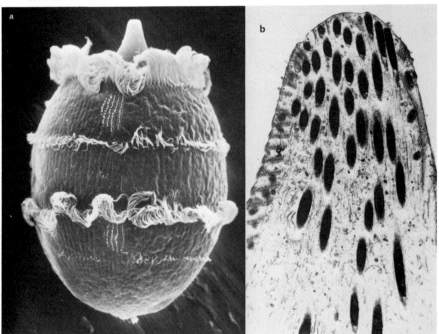

Fig. 8.7 Electron micrographs of *Didinium nasutum*. a, Scanning electron micrograph of the whole ciliate at an early stage in the preparations for division; the two prominent ciliated bands are present throughout the inter-division period, and the other two bands will develop further before the equatorial division of the cell. Micrograph by E. B. Small and D. S. Marszalek.[242] b, Transmission electron micrograph of a longitudinal section through the proboscis showing the bands of fibrils at the sides and giant trichocysts at the centre (unpublished).

When the cilia around the mouth are compounded together to form larger functional units, the invagination in which they lie, or the depression from which they originate, is called a buccal cavity. Such a buccal cavity is seen clearly in *Tetrahymena* (Fig. 1.4, p. 6), whose oral ciliature consists of three membranelles at the left side of the buccal cavity and an undulating membrane at the right side. Each membranelle is a triangular

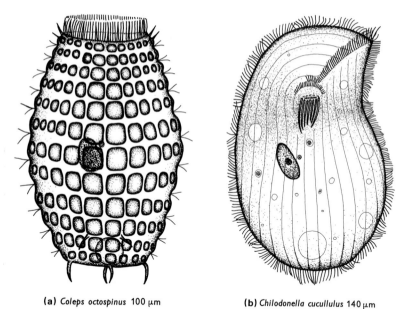

(a) *Coleps octospinus* 100 μm   (b) *Chilodonella cucullulus* 140 μm

Fig. 8.8 Rhabdophorine (a) and cyrtophorine (b) gymnostomes of two common genera.

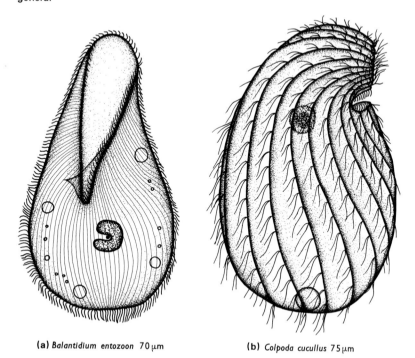

(a) *Balantidium entozoon* 70 μm   (b) *Colpoda cucullus* 75 μm

Fig. 8.9 Two common trichostome ciliates, (a) from the rectum of the frog and (b) from soil.

platelet compounded from 12 to 60 cilia arranged in three rows, and the undulating membrane is a single row of cilia borne on kinetosomes of the outer of two close set rows of basal bodies[76] (Fig. 8.11). These two characteristic types of compound ciliary structure, the undulating membrane and the membranelles, which are often extended to form a longer adoral zone of membranelles, occur in more or less modified form in many other orders of ciliates. They first appear in unspecialized members

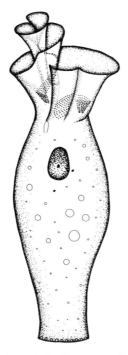

Fig. 8.10 The chonotrich ciliate *Spirochona gemmipara* (100 μm) from gills of *Gammarus*.

of the order **Hymenostomatida**, such as *Tetrahymena*, and these forms are thus held to represent an important stage in evolution. The wall of the buccal cavity of *Tetrahymena* is supported at the right side by a series of pellicular ribs underlain by rows of microtubules which are believed to originate from the barren kinetosomes of the undulating membrane system. These microtubules run together posteriorly to form a large post-oral fibre. Some microtubule ribbons from the membranelles run in the pellicle at the left side of the buccal cavity and are also thought to join in this post-oral fibre which extends alongside the cytopharynx into the endoplasm. A filamentous reticulum underlies the pellicle of the buccal cavity, and fibrils also link the membranelles with each other and with the undulating membrane. The somatic ciliature of hymenostomes normally consists of a complete covering of kinety rows, the cilia of which may be sparse or closely set.

The peniculine hymenostomes such as *Paramecium* probably represent an offshoot from the main line of evolution of ciliates. The body surface

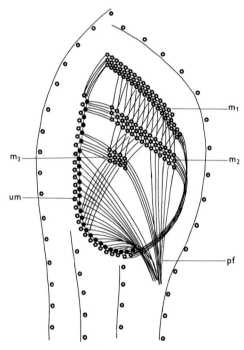

Fig. 8.11 The arrangement of kinetosomes and associated fibrils in the oral region of *Tetrahymena*. Kinetosomes of the three membranelles (m_1, m_2, m_3) and the undulating membrane (*um*) lie in the buccal cavity and are surrounded by somatic kineties; some root fibrils interconnect the compound ciliary structures and others run down into the cytoplasm as a post-oral fibre (pf). The barren kinetosomes of the undulating membrane are shown as filled circles[76] (redrawn).

at one side near the anterior end of *Paramecium* is depressed to form a vestibulum, within which is the opening to the deep, tubular, buccal cavity[129] (Fig. 8.12). Around the rim on the right side of the opening to the buccal cavity is a double row of kinetosomes, the outer of which bear the cilia that form the endoral membrane (equivalent to the undulating membrane of *Tetrahymena*). At the left side of the buccal cavity

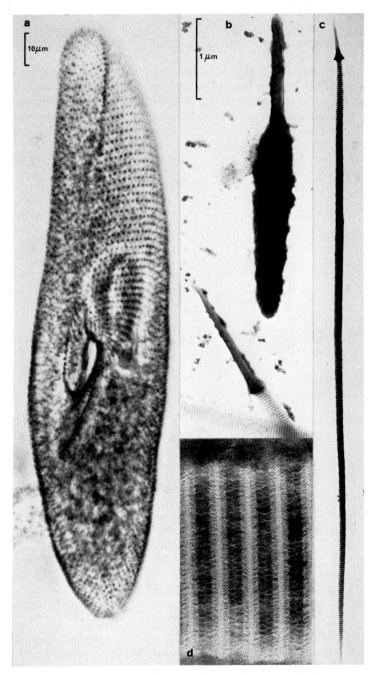

Fig. 8.12 a, Photomicrograph of *Paramecium caudatum* that has been stained with silver to show the arrangement of kinetosomes; the entrance to the buccal cavity is seen at the centre of the organism. Electron micrographs of trichocysts of *Paramecium*, b, before discharge, c and d (detail) after discharge. Electron micrographs by L. H. Bannister (unpublished).

are three groups of cilia, each composed of four rows, extending from the rim of the opening inwards to the posterior end of the buccal cavity (Fig. 8.13). These three groups are called, from right to left, the quadrulus, the dorsal peniculus, and the ventral peniculus; they are held to correspond with the three membranelles of *Tetrahymena*. The wall of the buccal

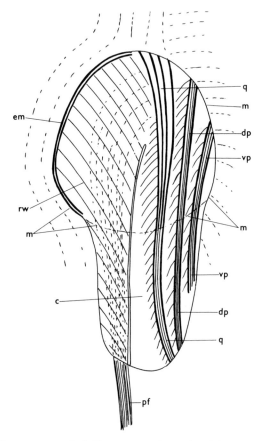

Fig. 8.13 The arrangement of ciliary rows and associated fibrils in the oral region of *Paramecium*. At the animal's right the buccal cavity is bordered by the endoral membrane (*em*), from which a ribbed wall (*rw*) extends to the cytostome (*c*), while at the left the buccal cavity contains three groups of four rows of cilia, the ventral peniculus (*vp*), the dorsal peniculus (*dp*) and the quadrulus (*q*), which run deep into the tubular buccal cavity; nemadesmal fibres (dotted) run dorsal to the ribbed wall and extend into the cytoplasm as the post-oral fibre (*pf*). Around the margin of the buccal cavity (*m*) the somatic ciliary rows line the wall of the vestibulum.

cavity at the right has a ribbed alveolate pellicle supported by one or two microtubules under each rib. Posteriorly, nemadesmal bundles of microtubules are seen beneath these ribs, and these bundles continue backwards as post-oral fibres around the cytopharynx. A ridged membrane also occurs in the dorsal wall of the buccal cavity between the cytostome and the quadrulus; at this side there are no alveoli, but the ridges are underlain by wide cytopharyngeal ribbons of microtubules, often with a second narrow ribbon beside each, and with numerous characteristic flattened vesicles between the ridges, possibly providing a source of membranes for food vacuole formation. These microtubules at the left side of the cytostome are derived from the kinetosomes of the quadrulus and extend back into the endoplasm posterior to the cytostome. The cytostome is a long triangular area covered by a single unit membrane and consisting of an irregular amoeboid surface. Much of the pellicle of the buccal cavity is underlain by a filamentous reticulum. Normal somatic ciliature is found in the vestibulum.

Another evolutionary branch within the Hymenostomatida is represented by the pleuronematine ciliates in which the undulating membrane is greatly expanded and in some cases forms a very prominent 'sail' at

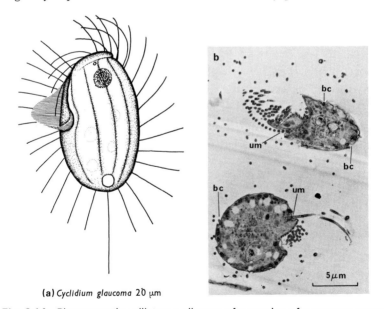

(a) *Cyclidium glaucoma* 20 μm

Fig. 8.14 Pleuronematine ciliates; a, diagram of a member of a common genus and b, an electron micrograph of sections at two different planes through an unidentified pleuronematine, showing cilia of the undulating membrane (*um*) and body cilia (*bc*).

the right side of the body, e.g. *Cyclidium* (Fig. 8.14). In these forms the three membranelles are less well developed; they are often narrow rows of cilia which lie close to the anterior part of the undulating membrane, in front of and perhaps even to the right of the cytostome, and usually one membranelle is much larger than the other two. It has recently been suggested that these ciliates should form part of a separate order, the Scuticociliatida.[241]

Similar modifications of the buccal ciliature are seen in some members of the order **Thigmotrichida**, most of which are found in the mantle cavities of molluscs, and are characterized by the possession of an area of 'thigmotactic' (adhesive) ciliature on the left side of the body. Oral ciliature is only present in the arhynchodine thigmotrichs, in which the cytostome is carried towards the posterior end in *Ancistrum* (Fig. 8.15a)

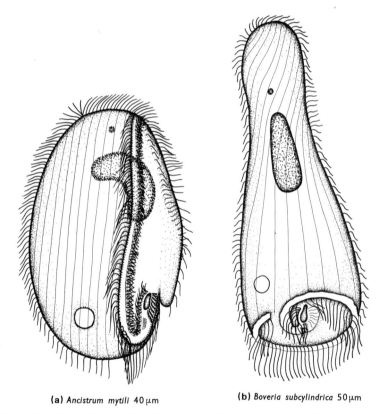

(a) *Ancistrum mytili* 40 μm   (b) *Boveria subcylindrica* 50 μm

Fig. 8.15 Two arhynchodine thigmotrich ciliates; a, from the gills of a lamelli-branch mollusc and b, from the respiratory tree of an echinoderm.

or to the posterior extremity in *Boveria*[161] (Fig. 8.15b). In these forms two 'kineties' of cilia are associated with the buccal cavity. At the right is the stomatogenous kinety or haplokinety, in which there is a double row of kinetosomes, only the outer row of which bear cilia, so that this kinety shows homologies with the undulating membrane. Immediately to the left of this is the adoral kinety, which runs closely parallel to the haplokinety until these two kineties separate at an area of ribbed pellicle just anterior to the cytostome. The adoral kinety is a polykinety two or

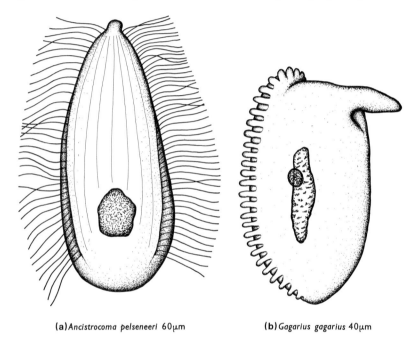

(a) *Ancistrocoma pelseneeri* 60μm   (b) *Gagarius gagarius* 40μm

Fig. 8.16 Two rhynchodine thigmotrich ciliates from the gills of the molluscs *Mya* (a) and *Mytilus* (b).

three kinetosomes wide, and seems to be the homologue of a peniculus, and therefore of a membranelle, but only a small fragment possibly representing one of the other two membranelles may be found. The rhynchodine (tentacle-bearing) thigmotrichs lack oral ciliature, and the area of thigmotactic cilia shows progressive reduction as the adhesive tentacle becomes more developed towards the ultimate sucker form.[208] In *Ancistrocoma* (Fig. 8.16a) the ciliature is fairly complete and the tentacle is supported by an internal cylinder of rows of microtubules reminiscent of that in suctorian tentacles,[134] while in *Gagarius* (Fig. 8.16b) the adult ciliate lacks

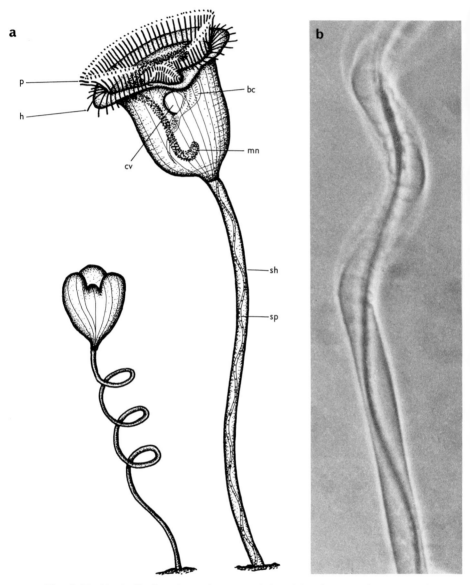

Fig. 8.17 *Vorticella*. Drawings of an extended zooid and a partially contracted one, and the photomicrograph at the right shows a portion of the stalk of a specimen which has almost completely extended following contraction although the manner in which the sheath of the stalk is distorted during contraction is still visible. *bc*, Buccal cavity: *cv*, contractile vacuole: *h*, haplokinety: *mn*, macronucleus: *p*, polykinety: *sh*, sheath of stalk and *sp*, spasmoneme (myoneme) of stalk.

cilia and the whole of one side of the body forms a sucker, although cilia are present on embryos that are formed by budding.

The buccal ciliature of ciliates of the order **Peritrichida** also consists of haplokinety and polykinety rows,[159] along both of which metachronal waves pass towards the cytostome. The cytostome lies at the base of a deep tubular buccal cavity, often called the infundibulum. Within the buccal cavity rows of cilia spiral upwards from the cytostome, making between $\frac{1}{2}$ and $2\frac{1}{2}$ turns before they reach the opening; outside the buccal cavity these rows continue to spiral in a clockwise direction, as seen from above, making from one half to four turns around the margin of the body

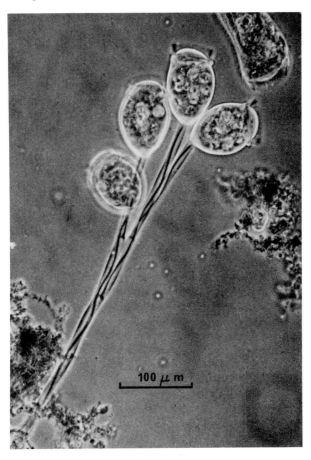

Fig. 8.18 Photomicrograph of a small colony of the peritrich *Carchesium*; the stalk is branched, but the spasmoneme of each branch is separate.

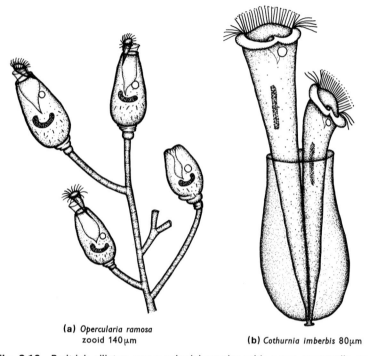

(a) *Opercularia ramosa*
zooid 140 μm

(b) *Cothurnia imberbis* 80 μm

Fig. 8.19 Peritrich ciliates, one a colonial species with a non-contractile stalk (a) and the other a loricate species (b) after division but before one of the daughters has migrated to form its own lorica.

at the oral end of the ciliate. The haplokinety remains throughout as a double row of basal bodies, accompanied in part by an extra inner row of barren kinetosomes; cilia are only present on the outer row of kinetosomes, and these form the outermost cilia of the buccal ciliature. Outside the buccal cavity, in the peristome region, the polykinety consists of a band 3 cilia wide running alongside the haplokinety, but within the buccal cavity two further bands of cilia, each 3 kinetosomes wide, normally appear between the original polykinety and the haplokinety; these triple rows of cilia resemble peniculi.

The Peritrichida probably form the largest order of ciliates, and *Vorticella* is probably the most familiar member of the group (Fig. 8.17). The suborder Sessilina contains solitary forms with stalks that are contractile (*Vorticella*) or non-contractile (*Rhabdostyla*), colonial forms with stalks that are contractile (*Carchesium*, Fig. 8.18) or non-contractile (*Opercularia*, Fig. 8.19a), loricate forms (*Cothurnia*, Fig. 8.19b) and stalkless forms which may be colonial (*Ophrydium*, Fig. 8.20). The majority of these forms are probably bacterial feeders (see p. 63). The contractile myoneme (spasmoneme) in the stalk of peritrichs has been the subject of intensive study;[8] extremely rapid shortening of the filamentous structure is excited by calcium ions, but direct participation of ATP is not required.[7, 127] Sessiline peritrichs may develop an aboral ciliary girdle, called a telotroch, from a ring of dormant kinetosomes following fission of the body or at other times when migration becomes necessary. At the extreme aboral end of the ciliate is a group of basal bodies that forms the scopula, a tuft of ciliary organelles involved in stalk formation. Body cilia are otherwise lacking, but the pellicular alveolar system still appears to be present.

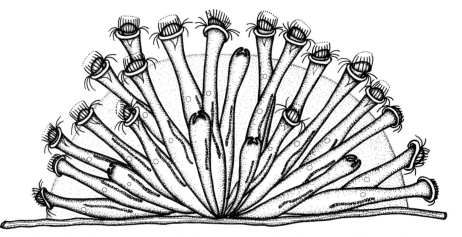

Fig. 8.20 The colonial peritrich *Ophrydium sessile* with many stalkless zooids (300 μm) in a gelatinous mass.

In the other suborder, the Mobilina, are forms which are motile and have a large adhesive basal disc with peripheral rows of cilia and often complex arrays of denticles. These forms are predominantly ectocommensals, and some are important parasites of fish, e.g. species of *Trichodina* (Fig. 8.21).

Some authors suggest that these ciliates represent the culmination of the evolutionary line hymenostomes → pleuronematines → thigmotrichs → peritrichs, while others more cautiously refer to common evolutionary trends finding expression in all of these groups of ciliates;[161] some impression of the similarity of oral ciliature may be gained from Fig. 8.22. In peritrichs the tendency to remain stationary for long periods in the stalked and loricate forms or for shorter periods in the mobiline forms could be taken as an extension of the thigmotactic habit.

A pattern of buccal ciliature similar to that in *Tetrahymena*, but in which the membranelles are increased in number to form a long adoral zone, is characteristic of the orders of 'spirotrich' ciliates. In such an adoral zone of membranelles the metachronal waves pass away from the cytostome along the row of compound cilia, which is contrary to the direction of metachronal waves along the rows of oral cilia in peritrichs. The least specialized of the spirotrichs comprise the order **Heterotrichida**, most of which retain a full covering of simple body cilia. Some of these ciliates have a large buccal cavity with a long undulating membrane as well as a long adoral zone of membranelles, e.g. *Condylostoma* (Fig. 8.23a), *Blepharisma*, but in many heterotrichs the undulating membrane is reduced or missing and the adoral zone of membranelles may run along the side

Fig. 8.21   The mobiline peritrich *Trichodina pediculus* found on *Hydra*.

of the body, e.g. *Spirostomum* (Fig. 8.23b), or around the oral end of the body, enclosing an area of the ciliated surface, as in *Stentor* (Fig. 8.23c). Each membranelle is composed of two or three rows of cilia, up to 20 or more in each row, and the kinetosomes of the membranelles give rise to nemadesmal fibres which run down into the cytoplasm and meet with fibres from other membranelles.[131, 210, 267] The undulating membrane appears to consist of two or three rows of cilia, and may not be entirely homologous with the undulating membrane of hymenostomes. Rootlet microtubules from the cilia of the undulating membrane may also make contact with the membranelle nemadesmal fibres by running behind the wall of the buccal cavity. At least part of the wall of the buccal cavity is

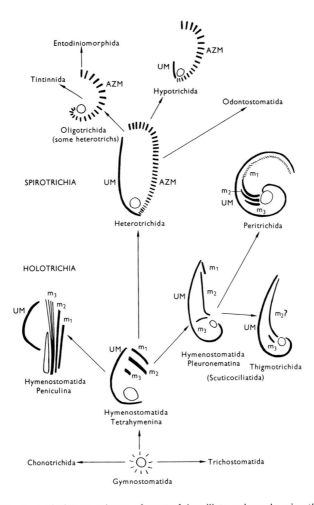

Fig. 8.22 An evolutionary scheme of most of the ciliate orders, showing the form of the buccal ciliature in representatives of the various groups. m_1, m_2, and m_3, Membranelles, *UM*, undulating membrane and *AZM* adoral zone of membranelles. Based on an evolutionary scheme by J. O. Corliss.[40]

underlain by ribbons of microtubules running back towards the cytostome. Stacks of rows of microtubular fibrils are associated with the body kineties, but the M fibres (p. 193) found in *Stentor* and *Spirostomum* are absent from *Blepharisma*, which is not contractile.

The body ciliature of other spirotrich ciliates may consist of few cilia

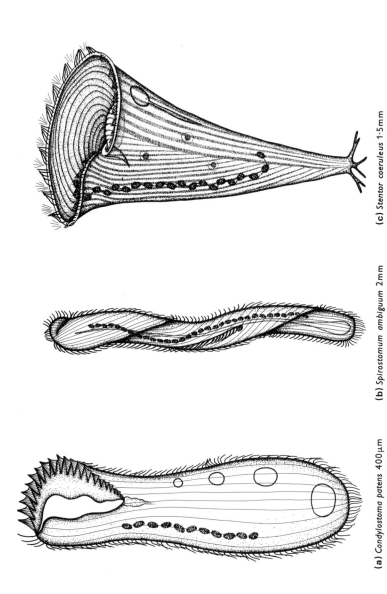

(a) *Condylostoma patens* 400μm   (b) *Spirostomum ambiguum* 2mm   (c) *Stentor coeruleus* 1·5mm

Fig. 8.23   Heterotrich ciliates of three common genera.

or cirri or may be absent, and the prominent adoral zone of membranelles is commonly used for locomotion as well as feeding. In the **Oligotrichida**, the majority of which are small marine forms, a small number of long body cilia are found on the common freshwater *Halteria* (Fig. 8.24a), while *Strombidium* (Fig. 8.24b) lacks somatic cilia. Members of the **Tintinnida** are pelagic forms, primarily from marine habitats, which secrete a gelatinous or pseudochitinous lorica (Fig. 8.25a), to which such foreign materials as sand grains or organic debris may be attached, as in *Codonella* and *Tintinnopsis* (Fig. 8.25b); these forms usually retain one

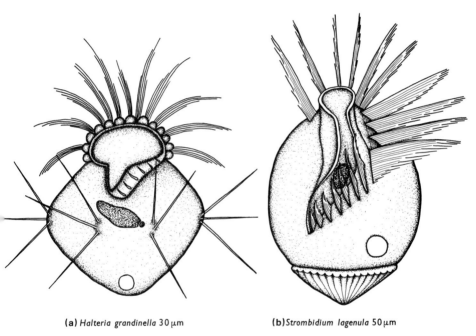

(a) *Halteria grandinella* 30 μm   (b) *Strombidium lagenula* 50 μm

Fig. 8.24   Oligotrich ciliates common in freshwater (a) and marine (b) habitats.

or many rows of body cilia, but swim by means of the adoral zone of membranelles which is protruded from the lorica. Ordinary somatic ciliature is absent in the **Entodiniomorphida**, which have a complex internal structure and a firm pellicle incorporating many microtubular fibrils, while the buccal ciliature of membranelles at the anterior end sometimes becomes separated into two or more clumps, and provides the origins of large bundles of nemadesmal fibrils. Members of this order are found in the digestive tract of herbivorous mammals, where they may be very abundant, e.g. *Entodinium* (Fig. 8.26a) from cows and sheep, and *Cyclo-*

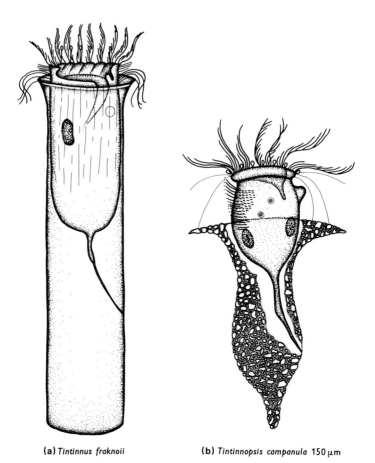

(a) *Tintinnus fraknoii*   **(b)** *Tintinnopsis campanula* 150 μm

Fig. 8.25   Tintinnid ciliates from marine (a) and freshwater (b) plankton.

posthium (Fig. 8.26b) from horses (p. 277). The **Odontostomatida** are a small group of laterally flattened ciliates with thick pellicles, often prominently ridged, with few body cilia and a reduced buccal ciliature of only about eight membranelles; they are found in fresh or salt water rich in organic materials, e.g. *Saprodinium* (Fig. 8.27).

Many of the most complex Protozoa are found in the order **Hypotrichida**. These are dorso-ventrally flattened ciliates with few dorsal cilia, and with ventral cilia formed into cirri which may form an extensive series of rows in *Urostyla* (Fig. 8.28a) and *Uroleptus*, marginal rows and scattered groups in *Stylonychia* (Fig. 8.28b), or only a small number of groups of cirri in *Euplotes* (Fig. 8.29a). In these forms the buccal cavity normally occupies the anterior region of the ventral surface, and the mouth ciliature sometimes includes a well developed undulating membrane as well as the prominent adoral zone of membranelles which usually extends along the anterior edge of the body. In one family of small discoid species, principally members of the common genus *Aspidisca* (Fig. 8.29b), there are no dorsal cilia, the membranelles are very small, and the cirri are reduced to anterior and posterior clumps. The component basal bodies of cirri give rise to large bundles of microtubular fibrils, most of which

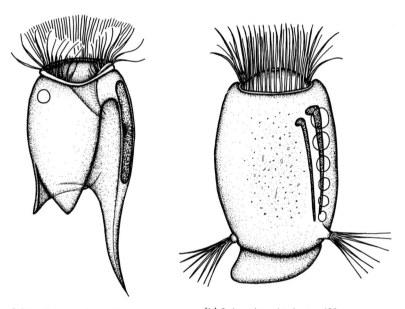

(a) *Entodinium caudatum* 70 μm   (b) *Cycloposthium bipalmatum* 100 μm

Fig. 8.26 Members of two common genera of entodiniomorph ciliates from the rumen of cows (a) and the large intestine of horses (b).

run in the pellicle, and whose function appears to be anchorage. The membranelles are arranged in a similar manner to those of heterotrichs (Fig. 8.30), although the nemadesmata seem to be less developed than pellicular fibrils in the mouth region of *Euplotes*.[87, 269] Hypotrichs show a characteristic backward jump, spontaneously or as an avoiding reaction; this is achieved by cirri, particularly the large cirri of the posterior anal (transverse) group, whose reversed beating is associated with depolarization of the cell membrane, and may be induced by mechanical stimulation of the anterior end of the body.[183]

The *Apostomatida* are a small order of ciliates which may have been derived from gymnostomes and which have complex life cycles, the morphogenetic stages of which were the subject of a classic study by Chatton and Lwoff in the 1930's.[162] The life cycle characteristically involves a feeding stage (trophont), during which the ciliate grows without dividing, and has few cilia in spiral rows and a rosette structure near the cytostome (Fig. 8.31). This is followed by a stage of division (tomont),

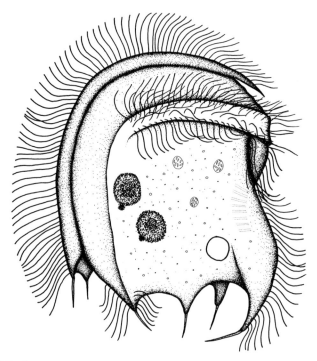

Fig. 8.27 *Saprodinium dentatum* 70 μm an odontostome ciliate.

usually in a cyst, during which a number of small ciliated tomites are produced; these tomites do not feed, but swim away and encyst, usually on a crustacean, to form a phoront within which the ciliate completes the transformation to the trophont form. In *Foettingeria* the tomite forms the phoront cyst on a crustacean, and hatches to release the trophont when the crustacean is eaten by an anthozoan coelenterate; the tomont is found on the outside of the anthozoan. In some other forms the trophont may grow in the cast skin of a crustacean, and the tomite seeks out a living crustacean, upon whose gills or integument it encysts to form a phoront; an example of this type is *Polyspira* (Fig. 8.32), found on the hermit crab *Eupagurus*.

A number of mouthless ciliates have been grouped together in the order *Astomatida*;[206] this is probably not a natural group of ciliates. They

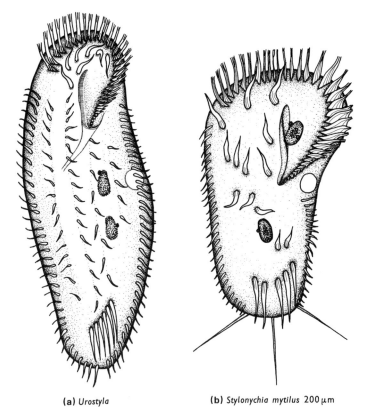

(a) *Urostyla*   (b) *Stylonychia mytilus* 200 μm

Fig. 8.28 Two hypotrich ciliates in which rows of cilia persist on the ventral surface in addition to groups of cirri and the adoral zone of membranelles.

generally have a uniform body ciliation and often specialized holdfast structures, suggesting a possible relationship with the thigmotrich ciliates. Binary fission of these ciliates may be incomplete, so that catenoid colonies are formed. Astome ciliates are parasites found in invertebrates, particularly annelid worms, e.g. *Anoplophrya* and *Metaradiophrya*, which are found in the intestine of earthworms and *Durchoniella* (Fig. 8.33) in a polychaete.

Members of the **Suctorida** are devoid of cilia during the mature, feeding stage; they are normally sessile and feed by means of tentacles through which organic materials, usually from the cytoplasm of other ciliates, are sucked in to form food vacuoles (p. 64). Suctorians reproduce

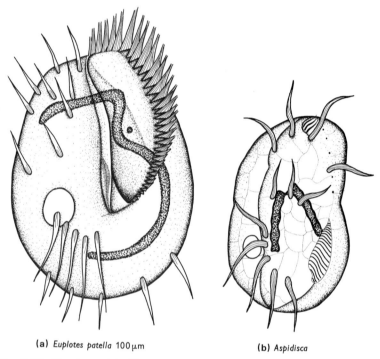

(a) *Euplotes patella* 100 μm   (b) *Aspidisca*

Fig. 8.29   Two hypotrich ciliates in which the body ciliature is reduced to groups of cirri. Both are representatives of common freshwater genera; *Euplotes* often has only four cirri at the caudal margin, and other species of *Aspidisca* are frequently more nearly circular in outline.

by budding off ciliated larvae, either exogenously or in an invagination, the cilia developing from dormant kinetosomes of the adult.[13] The larvae swim away and in most forms a new stalk develops from a ciliary scopula. These protozoans are undoubtedly ciliates since they show the characteristic nuclear dimorphism and the sexual process of conjugation as well as a ciliated larval stage. Such stalked forms as *Podophrya* (Fig. 8.34) and *Acineta* (Fig. 8.35a) are common, while the gills of freshwater species

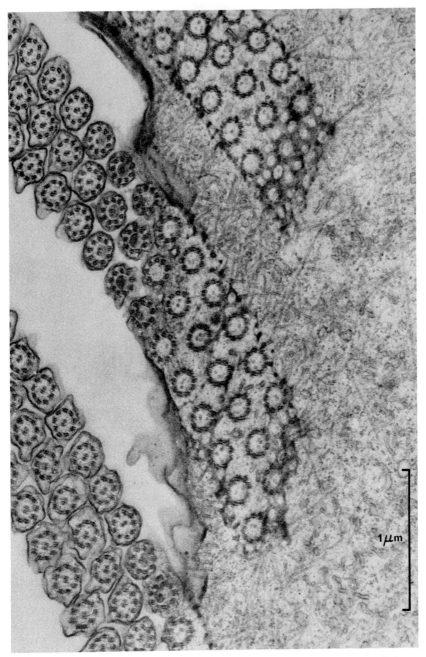

1 µm

Fig. 8.30 Electron micrograph of a section through the basal region of the membranelles of *Euplotes eurystomus*. Micrograph by R. Gliddon (unpublished).

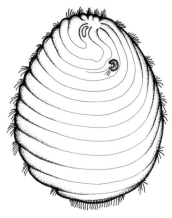

Fig. 8.31 The trophont stage of the apostome ciliate *Foettingeria actiniarum* (up to 1 mm long) from the gastrovascular cavity of sea anemones.

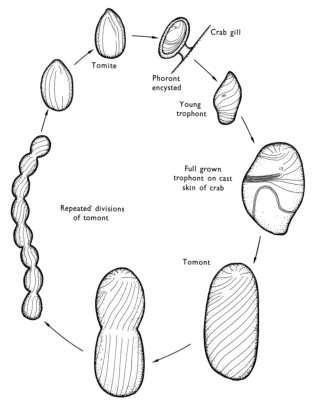

Fig. 8.32 The life cycle of the apostome ciliate *Polyspira delagei*, found as an encysted phoront on the gills, or as a trophont on the cast skin, of the crab *Eupagurus bernhardus*. Redrawn from Lwoff.[162]

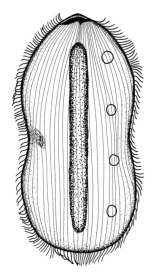

Fig. 8.33 The astome ciliate *Durchoniella brasili* 150 μm from the polychaete worm *Audouinia*, showing the anterior skeletal bar which is developed into a prominent hook in other genera.

of the crustacean *Gammarus* frequently carry *Dendrocometes* (see p. 267), which has no stalk, but whose numerous branched arms have suctorial tips. These forms feed on swimming ciliates, but *Choanophrya* has been

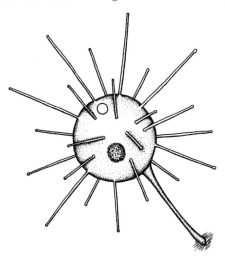

Fig. 8.34 The suctorian ciliate *Podophyra collini* 50 μm.

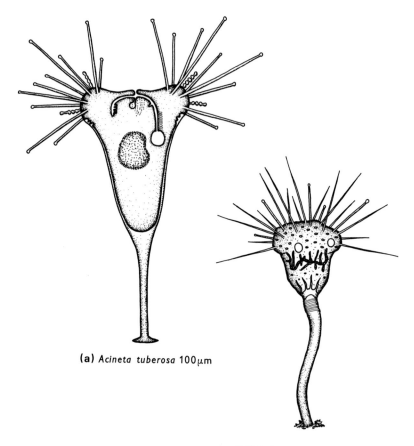

(a) *Acineta tuberosa* 100μm

(b) *Ephelota gemmipara*, body 200μm

Fig. 8.35 Two suctorian ciliates. *Acineta* has tentacles of a single type and the drawing shows an early stage in the formation of a larva within a brood chamber by endogenous budding (information from Bardele).[13] In *Ephelota* there are pointed prehensile tentacles and capitate suctorial tentacles; this genus is marine.

observed to feed on tissues of dead animals, and *Tachyblaston* is parasitic on the suctorian *Ephelota*. The tentacles of suctorians are sometimes rigid and in other cases are retractile; in *Ephelota* (Fig. 8.35b) pointed pre-hensile tentacles occur among the knobbed suctorial tentacles, and in-ternally the microtubular cylinder characteristic of feeding tentacles is replaced by an axial bundle of microtubules in the prehensile tentacle (Fig. 8.36).

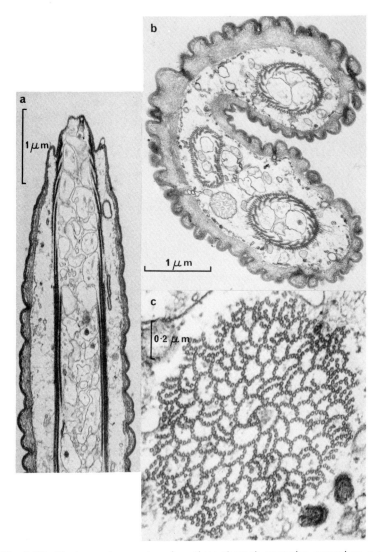

Fig. 8.36 Electron micrographs of sections through suctorian tentacles; *a* a longitudinal section through the tip of a tentacle of *Dendrocometes paradoxus*, showing the microtubule cylinder and a haptocyst at the apex:[15] *b* a transverse section through a main branch of a tentacle of *Dendrocometes* showing three cylinders of microtubules:[15] *c* a transverse section through the microtubular axis of a prehensile tentacle of *Ephelota* (unpublished). Micrographs by C. F. Bardele.

9

The Sub-Phylum Sporozoa (=Apicomplexa)

All members of this sub-phylum are endo-parasites with complex life cycles. They have long been regarded as distinct from other Protozoa because of the presence of spores containing the infective sporozoites, hence the name. However, in a number of groups the sporozoites are naked, and the possession of these infective individuals, which are elongate cells that are not amoeboid and move by gliding or body flexion, is more characteristic than the possession of spores. Recent studies of fine structure have revealed several features that appear to be better diagnostic characteristics of the group;[277] these features occur widely in all major groups of Sporozoa, but the full extent of their distribution is not known. In at least one stage of the life cycle the parasite possesses a complex of polar organelles comprising a conoid, apical rings and dense thread-like bodies which are called by various names including toxonemes, sarconemes, rhoptries, micronemes and paired organelles (Fig. 9.1). The conoid is apparently formed from a spiral band, and may be compressible and extensible, while the dense bodies may contain material (enzymes?) for release outside the cell. Most or all of these structures are normally present in sporozoites, whose sub-pellicular microtubules terminate anteriorly at the posterior apical ring. The complex also occurs in some trophozoites, in merozoites (see below) and in other stages of at least some species. Even more widely distributed through the life cycle stages of Sporozoa are characteristic micropore structures appearing as short dense cylinders beneath the cell surface. Frequently these parasites are surrounded by three membranes, and only one of these covers the micropore.[278] It has been suggested that the micropores are cytostomes or sites of extrusion of material from the cell; in some species there are many micropores

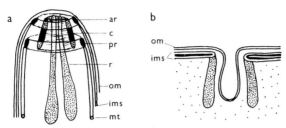

Fig. 9.1 Some characteristic structures of sporozoans; *a*, polar organelles—anterior apical ring (ar), conoid (c), posterior apical ring (pr), rhoptries (r), outer membrane (om), inner membranes (ims) and microtubules (mt); *b*, a micropore, with outer membrane invaginated within the cylindrical dense collar and inner membranes interrupted at the outer end of the collar.

distributed over the cell surface, and additional comparable structures have been found in the pellicle of larger sporozoan cells.

The presence of the polar ring structure as an apical complex in all members of the group that have been studied has led to the recent proposal by N. D. Levine that the name of the sub-phylum should be changed from Sporozoa to Apicomplexa (see p. 283), and this name is now being used quite widely. The sub-phylum is divided into two classes, the Telosporea and the Piroplasmea.

CLASS TELOSPOREA[90, 92, 165]

Generalized life cycle

The life cycle of all telosporeans shows several characteristic stages, the sequence of which is shown in Fig. 9.2. The parasite makes its entry into the host tissues as a sporozoite. These small infective stages, which do not have flagella or pseudopodia, but show flexural and gliding move-

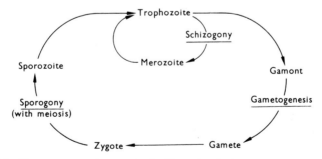

Fig. 9.2 The sequence of stages in the life cycle of telosporean sporozoans.

ments, enter the cells of the appropriate tissue of the host, perhaps with the aid of the polar organelles. Within the host cells the Protozoa enter a phase of growth and are known as trophozoites. Parasites may go through one or more stages of asexual multiple fission (schizogony) in the host cells, producing numerous merozoites which enter other cells and grow there, or may leave the host cells and continue to grow extracellularly. Eventually the trophozoites transform to gamonts which differentiate to form gametes. The zygotes which arise by fusion of the gametes undergo meiosis (p. 81) during the first of a series of subsequent divisions that result in the formation of a number of sporozoites. The sporozoites of many species are transmitted from one host to another in resistant spores, but naked sporozoites may be transmitted directly to new hosts by vectors. The class Telosporea is divided into two sub-classes, the Gregarinia and the Coccidia; in the sub-class Gregarinia the trophozoites grow large in extracellular body spaces, and in the sub-class Coccidia the trophozoites are small and remain within cells.

The sub-class Gregarinia

Various species of *Monocystis* and of several related genera are common parasites of earthworms, where examples may usually be found in the seminal vesicles. The normal life cycle is shown in Fig. 9.3. It is assumed that the earthworms eat the sporocysts of *Monocystis* and that the sporozoites escape in the gut and find their way to the seminal vesicles. Here the parasite enters the central cell of a sperm morula and proceeds to grow. The trophozoite grows too large for the host cell and then lives free in the seminal fluid, where growth is completed. Two full-grown trophozoites associate in pairs as gamonts, and a common gamontocyst wall is secreted around them. Large numbers of amoeboid gametes are formed by each gamont, and these subsequently fuse in pairs within the gamontocyst. Following the secretion of a sporocyst wall around each zygote, these diploid cells divide three times forming eight haploid sporozoites within each sporocyst. Gamontocysts containing numerous sporocysts are frequently found in the seminal vesicles of earthworms, but it is not clear how these find their way to the soil to infect other worms, except at the death of the host.

Other gregarines may commonly be found in the body cavities and gut of other worms and in insects. A good source of material for laboratory study is the gut of larvae of the meal-worm (*Tenebrio*), where several species of *Gregarina* may be found. In these forms the body has three parts, the epimerite by which the young parasite anchors itself to the host cell, a clearer protomerite region and the more dense deutomerite region which contains the nucleus. Older trophozoites lose their attach-

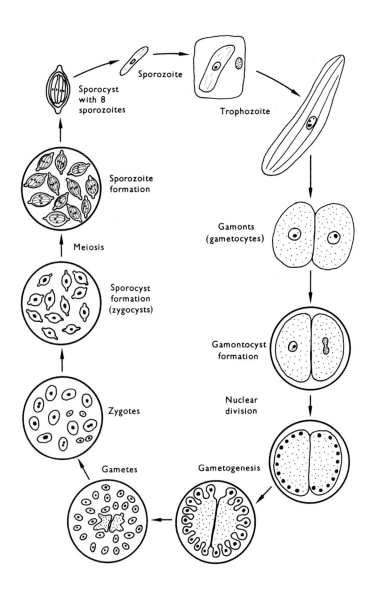

Fig. 9.3 Stages in the life cycle of the gregarine *Monocystis* (see text).

ment to host cells and associate in pairs (in syzygy) at an early stage in
growth (Fig. 9.4b). The first layers of the gamontocyst are not formed
until growth is completed, and the cysts are passed out with the faeces
before gametogenesis and spore formation have taken place inside them.

Increase in numbers of eugregarines like *Monocystis* and *Gregarina* only
occurs in the formation of gametes and spores within the gamontocyst.
There are a few species that are classified with gregarines because they
have large extracellular trophozoite stages, but which show phases of
increase in numbers by schizogony during the intracellular phase, as well

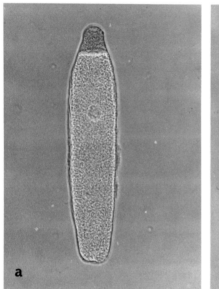 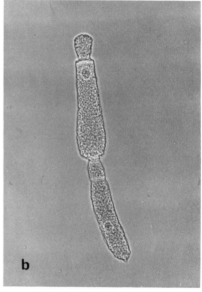

Fig. 9.4 Photomicrographs of trophozoites of *Gregarina* from the gut of a meal-
worm; a, a single trophozoite showing the division of the body into protomerite
and deutomerite regions; b, two trophozoites attached in syzygy.

as division within the gamontocyst. These forms may produce one or more
generations of merozoites before leaving the host cells as trophozoites and
proceeding to perform the characteristic gregarine gamontogamy. Because
of the occurrence of schizogony these species may build up much larger
infections within the host than may be found in *Gregarina* or *Monocystis*.
Such a life cycle has been found in some species of *Selenidium*, which
are common archigregarine parasites in the gut of polychaetes and a few
other marine invertebrates, but in other species of the genus no traces
of schizogony have yet been found.[227] The trophozoites of *Selenidium*

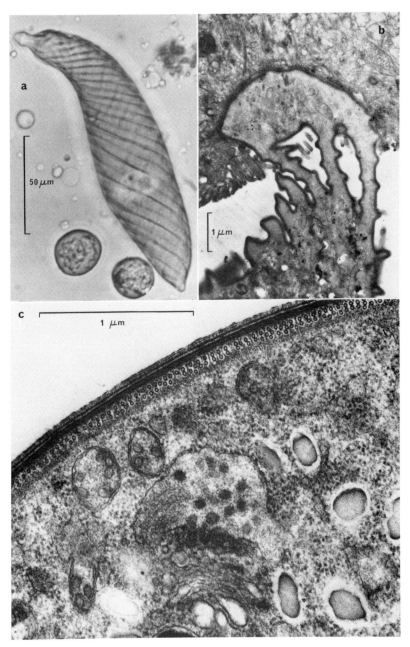

Fig. 9.5 *Selenidium terebellae. a,* Photomicrograph of a trophozoite from the gut of the worm *Terebella lapidaria; b,* electron micrograph showing the attachment of the anterior end of a trophozoite to the gut wall of the host; *c,* electron micrograph of a transverse section through a trophozoite showing the three membrane layers underlain by a layer of microtubules in an epiplasmic zone and mitochondria and a dictyosome in the cytoplasm. Micrographs by A. G. H. Dorey (unpublished).

(Fig. 9.5) are large vermiform cells with a striated pellicle which is re-inforced by large numbers of microtubular fibrils that are probably involved in the undulatory movements of these organisms.[228] These trophozoites may remain attached to cells of the gut epithelium by the anterior tip of the body. Schizogony also occurs in *Mattesia* and some related neo-gregarines of insects.

Sub-class Coccidia: tissue parasites[54]

The sub-class Coccidia contains two orders; the order Protococcida contains only two species parasitic in marine worms, while the important coccidian species are in the order Eucoccida, which is divided into three sub-orders, the Eimeriina and Adeleina, normally found as tissue parasites, and the Haemosporina, found in blood cells. Members of the first two

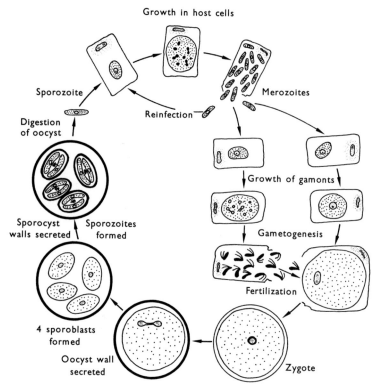

Fig. 9.6 Stages in the life cycle of the coccidian *Eimeria steidae,* a parasite in the liver cells of rabbits (see text).

sub-orders normally mature within a single host, which they enter through the mouth as oocysts and which they leave as oocysts in the faeces or urine.

The life cycle of *Eimeria steidae* (sub-order Eimeriina), found in the bile ducts within the liver of rabbits, will be described as an example (Fig. 9.6). The oocyst ingested by the rabbit hatches in the small intestine of the host, where the cyst walls are probably digested by trypsin. The sporozoites which emerge enter the gut wall and pass via the hepatic portal blood system to the liver where they enter the epithelial cells of the bile ductules. Within these cells the parasite grows as a trophozoite before undergoing schizogony to produce numerous merozoites which break out of the host cell and enter other host cells to undergo another cycle of schizogony. After several cycles of schizogony the trophozoites develop into gamonts instead of schizonts. Two forms of gamont can be distinguished: macrogamonts which form a single macrogamete that remains within the host cell, and microgamonts which divide to produce numerous microgametes, each with two free flagella and one recurrent attached flagellum, that swim away to fuse with macrogametes. The zygote secretes a thick oocyst wall, divides by a one-division meiosis (p. 73) and by mitosis to form four sporoblast cells, each of which secretes a sporocyst wall and divides once more to form two sporozoites. The oocysts leave the host cells early in sporogony and this process is completed outside the host within a few days of the cysts being passed out with the faeces. Contamination of food is presumed to be the means of infection of new hosts, and the coprophilous habits of rabbits make infection more likely. This parasite causes severe hepatitis of the rabbits and is sometimes fatal.

Other species of *Eimeria* are pathogenic in various domestic animals,[9] including *E. necatrix* and *E. tenella*, which are found in the small intestine and caeca of chickens and are often fatal, species causing diseases in ducks and geese, and several species which cause various forms of 'coccidiosis' in cattle. Oocysts of the related genus *Isospora* can be distinguished from those of *Eimeria* because species of *Isospora* are characterized by the presence of four sporozoites in each sporocyst and two sporocysts per oocyst (Fig. 9.7); similarly, other genera of the family Eimeriidae may also be distinguished by the number and arrangement of sporozoites in the oocyst. Several species of *Isospora* are found in domestic animals, including the pathogenic *I. bigemina*, which is found in the small intestine of dogs. Related species are occasionally found in man, but are apparently only mildly pathogenic.

Most species of *Eimeria* and *Isospora* are quite specific in their hosts; by contrast, *Toxoplasma gondii* has been found in a very wide range of birds and mammals, including man, from all parts of the world. The

Fig. 9.7 Oocysts of two common genera of coccidians, *a, Eimeria* (35 × 20 μm) and *b, Isospora* (30 × 15 μm) (see text).

rate of infection with this parasite is estimated to exceed 25% in man and to exceed 50% in some domestic animals in Britain and the U.S.A., and it may reach even higher levels in some other countries. The parasites appear in macrophage cells, where they multiply, apparently by a special form of binary fission, until they eventually fill the cell and burst it. Repeated cycles of this activity produce enormous numbers of small parasitic cells. Some of these 'zoites' find their way to other tissues—nerve, liver, kidney, muscle and lung—where they form resistant cysts containing numerous *Toxoplasma* cells (Fig. 9.8b). For many years the

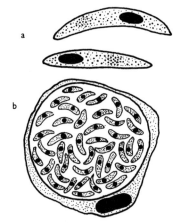

Fig. 9.8 *Toxoplasma gondii*; a, two isolated zoites (5 μm long) and b, cyst within a host cell, which may be reduced to a wall around the parasites.

relationships of these forms were obscure and they were classified in a separate class. It has recently been discovered that infected animals produce characteristic oocysts in the faeces, and careful study of the disease in cats has revealed that ingestion of *Toxoplasma* cysts is followed by profuse schizogony and gametogony of a coccidian type in the epithelium of the small intestine.[71, 118] These observations led to the conclusion that *Toxoplasma* is related to *Isospora*, but is unusual in forming cysts containing trophic cells, since cysts of other Sporozoa are formed following zygote formation. Toxoplasmosis is normally a mild disease in human adults, but is a more severe pre-natal disease in which the embryo is presumably infected through the placenta; damage caused by the parasite to the brain, the retina and the liver frequently causes death, blindness or jaundice.

A parasite with some similar ultrastructural features is *Sarcocystis*, which is found encysted in the muscles of warm-blooded vertebrates, including man. The cysts may be 1 or 2 mm long, and subdivided by trabeculae between which are numerous protozoans about 5 to 15 μm

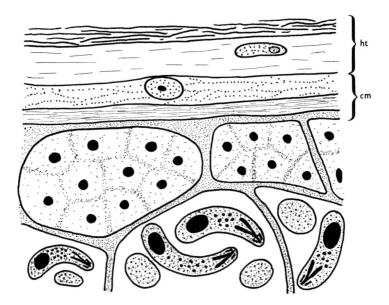

Fig. 9.9 *Sarcocystis tenella* from sheep muscle. A portion of the cyst is shown surrounded by connective tissue of the host (ht) and enclosed within a three layered cyst membrane (cm), the innermost layer of which is continuous with trabeculae which isolate compartments containing trophozoites of *Sarcocystis* at different stages of growth—some are immature and of irregular shapes, while others are crescentic with prominent polar organelles and about 10 μm long.

long (Fig. 9.9). Very little is known about this parasite, which is extremely common in sheep and cattle, but is not normally pathogenic.

The life cycles of other coccidian parasites differ in a variety of ways from that described for *Eimeria*. The parasite *Aggregata* (sub-order Eimeriina) is exceptional in that the life cycle takes place in two hosts. Schizogony occurs in a crustacean, producing many small parasite cells. These remain dormant until the host is eaten by a cephalopod mollusc, but then they become active and grow in the gut tissues to form gamonts and gametes. After fertilization sporogony takes place to produce numerous resistant spores, each containing three sporozoites. The spores are passed out with the faeces and may be eaten by a crustacean to start a new cycle. In members of the sub-order Adeleina, gamonts associate together before gametogony. *Klossia* is an adeleid parasite found in the kidney of gastropod molluscs, including terrestrial snails and slugs. It undergoes schizogony, gamont formation, mating of gamonts, gamete formation, fertilization and sporogony in the epithelium of the kidney, and oocysts containing sporozoites in thin walled sporocyst envelopes are passed out in the urine. A new host is infected by ingestion of cysts.

Sub-class Coccidia: blood cell parasites

Parasites of the genus *Plasmodium* have a life cycle which involves sexual reproduction and sporogony in mosquitoes and asexual reproduction (schizogony) in vertebrates. These malaria parasites, which belong to the sub-order Haemosporina have been the subject of a recent detailed monograph by P. C. C. Garnham,[82] and their fine structure has been reviewed by M. Rudzinska.[218] Malaria parasites enter the erythrocytes of the vertebrate and metabolize haemoglobin with the production of pigment masses within the cells of the parasite; other parasites that enter erythrocytes do not contain these pigment masses.

Several species of *Plasmodium* are found in man, all with a similar life cycle (Fig. 9.10). Naked fusiform sporozoites are injected into the skin with the saliva of an infected mosquito (*Anopheles*, Diptera) as it commences to feed. The sporozoites pass in the blood to the liver, where they enter parenchyma cells and grow. After between 5 and 15 days they undergo a stage of schizogony, known as preerythrocytic schizogony, which results in the production of 10–30 000 merozoites. Upon emergence these merozoites normally enter erythrocytes, although some merozoites may enter liver cells and go through one or more stages of exoerythrocytic schizogony. Trophozoites that enter erythrocytes are first seen as a vacuolated ring stage, and subsequently grow to fill much of the cell before commencing schizogony. Between 6 and 24 merozoites are released at the breakdown of the erythrocyte; these enter other erythrocytes and repeat the growth and schizogony cycle. Some of the trophozoites become gamonts instead of schizonts, and differentiate to form macrogamonts and microgamonts within the erythrocytes. These gamonts remain dormant in the erythrocyte unless this is ingested by a mosquito. Development

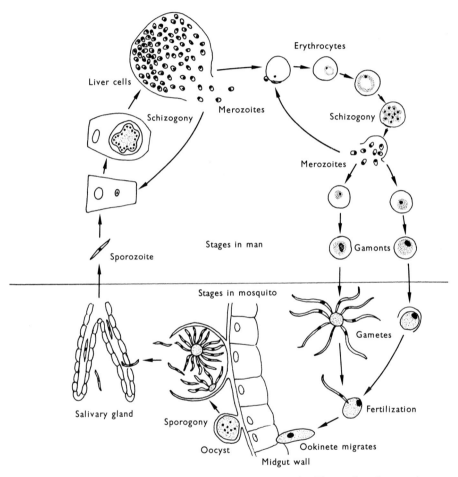

Fig. 9.10 Stages in the life cycle of the malaria parasite *Plasmodium* (see text).

of the parasites quickly recommences in the stomach of the insect and both forms of gamont emerge from the red cells. The macrogamont becomes a spherical macrogamete without division. The microgamont undergoes three nuclear divisions, and develops eight flagellar projections;

one nucleus enters each projection, and this then breaks away as a motile microgamete. Fertilization takes place in the stomach of the mosquito. The macrogametes are transformed after fertilization into an active zygote (ookinete), which burrows through the walls of the stomach and comes to lie on the outer surface of the stomach, where it secretes a thin oocyst wall. Growth of the zygote is followed by sporogony, which results in the production of very large numbers of sporozoites within each oocyst. The sporozoites break out of the oocyst into the haemocoel of the insect and many of them pass into the salivary glands, whence they may be injected into a new vertebrate host at the next feed of the mosquito.

The most pathogenic species of malaria parasite found in man is *P. falciparum*, which probably causes more human deaths in tropical countries than any other disease organism. The other species, *P. vivax*, *P. malariae* and *P. ovale*, do not usually cause fatal disease, although they do cause debilitating recurrent fevers. When the erythrocytes burst and the merozoites are released, other materials are liberated into the blood and it is thought that some of these are the toxic elements that cause fever. Schizogony tends to become synchronized so that vast numbers of erythrocytes burst at the same time, giving rise to fever every 48 hours in three of the species, or every 72 hours in *P. malariae*. The erythrocytes containing parasites may normally be found circulating freely in the blood, but in *P. falciparum* only the ring stages and mature gamonts are common in blood smears, since erythrocytes containing trophozoites of this species tend to stick in the blood capillaries. Here they cause capillary blockage, resulting in oxygen shortage in tissues and often rupture of the blood vessels. If capillary blockage occurs in the brain, death usually follows, and if it occurs elsewhere it may also do serious damage. The production of antibodies by the host,[252] or treatment with drugs,[214] can enable the host to destroy the erythrocytic stages, but in *P. vivax* and *P. malariae* the disease organism may persist over many years in exoerythrocytic sites, probably as dormant schizonts in the liver, from which it may emerge repeatedly to cause further outbursts of fever ('relapses').

Other species of *Plasmodium* are found in other mammals, birds and reptiles. Two species that have been widely studied because they can be maintained in laboratory animals are *P. berghei* in rats and mice and *P. gallinaceum* in chickens; they are used among other things for the testing of anti-malarial drugs. Species of the family Haemoproteidae, which are common in mammals, birds and reptiles, do not have an erythrocytic schizogony stage, so that the only parasites found within red blood corpuscles are the gamonts. Schizogony takes place in various visceral organs, such as liver and lung, and the parasites are transmitted by midges and other dipterans. A further family of haemosporian parasites contains *Leucocytozoon*, which is common in domestic and wild birds, and is transmitted in known cases by the dipteran *Simulium*; again only the gamonts occur in blood cells, and in this case the blood cells may become greatly enlarged.

Some members of the sub-order Adeleina also inhabit blood cells, but are distinguished from members of the sub-order Haemosporina because in the former the gamonts associate together before gametogony and because pigment masses are not found in parasites within erythrocytes. *Haemogregarina* and related adeleid blood parasites undergo schizogony in red or white blood cells or other cells of vertebrates, and the gamonts grow in blood cells. If a blood-sucking invertebrate—insect, leech, tick or mite—of the appropriate species sucks blood containing the parasites, the gamonts associate in pairs in the gut of the invertebrate, form gametes, perform fertilization and form a motile zygote. After migration to the appropriate part of the body the zygote encysts and undergoes sporogony to form numerous sporozoites. Some of these may find their way back to the vertebrate host, usually by being ingested by the vertebrate rather than being injected by the invertebrate.

Members of the genus *Lankesterella* are blood parasites of the sub-order Eimeriina. They are found in birds, reptiles and amphibia, and differ from other blood parasites in that both schizogony and sporogony occur in tissue cells of the vertebrate host. The resulting sporozoites enter blood cells which are ingested by an invertebrate vector and passed without development to another vertebrate host if that vertebrate eats the vector.

CLASS PIROPLASMEA[9]

These small pear-shaped parasites are found in the erythrocytes of vertebrates; they do not contain pigment masses formed from haemoglobin, do not form spores, are surrounded by only a single membrane[279] (Fig. 9.11), and do not appear to reproduce sexually; ticks are the only known vectors. Piroplasms have been found in vertebrates of all classes, and occasional infections have been recorded in man, but only in individuals who had previously had their spleen removed. Several species of *Babesia* are pathogenic in cattle, causing diseases such as red-water fever, and other important diseases are caused by members of this genus in horses and dogs. These diseases, and those caused by the related *Theileria*, are probably the most important diseases of domestic animals throughout much of the world.

In the vertebrate host, *Babesia* is only found in erythrocytes, where the parasites appear as small organisms 1–5 μm across (Fig. 9.12), which divide to produce two or four merozoites. These break out of the erythrocytes, releasing toxic products which cause fever and the excretion of blood pigments; hence the name 'red-water fever'. Infected animals may show anaemia and fever, accompanied by blockage of capillaries and anoxia in the tissues, as well as a variety of other manifestations of disease. The discovery in 1893 by Smith and Kilborne of the transmission of

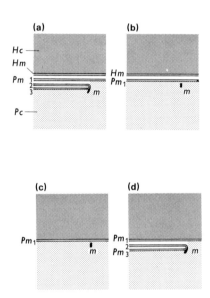

Fig. 9.11 Diagrams to show the number and type of membranes occurring between the host cytoplasm and parasite cytoplasm in various Sporozoa; a, *Toxoplasma* and other coccidians; b, trophozoite of *Plasmodium*; c, trophozoite of *Babesia*; d, merozoite of *Babesia*. *Hc* is the host cytoplasm, *Hm* a membrane of host origin, Pm_1 is the plasma membrane of the parasite, Pm_2 and Pm_3 are inner membranes of the parasite, *Pc* is parasite cytoplasm and *m* are micropores. Information mainly from papers by E. Vivier et al.[278,279]

red-water fever by ticks was an important event in parasitology, since this was the first demonstration of the transmission of a disease by an arthropod vector. Even now, however, the life cycle of *Babesia* in the tick is not clearly understood, and, despite some reports to the contrary, the occurrence of a sexual stage is not established. Ticks take few, large, blood meals, and they may only feed once in each instar of growth. Parasites which are ingested by a tick larva (or nymph) undergo a period of multiplication in the tick[79] and survive the moult of the host to be injected with saliva at the next feed. After multiplication in an adult tick the parasites may enter the eggs of the tick, and survive there throughout development to be injected with the saliva of the tick larva at its first meal. *Babesia* species of veterinary importance occur in all continents of the world.

   Theileria occurs in erythrocytes as small trophozoites similar to those of *Babesia*, but the two genera differ in that exoerythrocytic schizogony

Fig. 9.12   Two cells of *Babesia bigemina* in an erythrocyte cell.

also occurs in *Theileria*. These exoerythrocytic schizonts are found in lymphocytes as larger, rounded bodies up to 20 μm across; the numerous merozoites released after schizogony invade erythrocytes. The transmission by ticks is probably similar to that of *Babesia*. Species of this genus cause serious diseases in cattle and sheep in Africa and Asia.

   Piroplasms also occur in cold-blooded vertebrates where they have been seen as trophozoites in erythrocytes, but their life cycle and mode of transmission are not known.

IO

The Sub-Phylum Cnidospora

These parasitic forms infect their host as small amoeboid individuals which emerge from ingested spores and multiply within the host. Eventually they differentiate to form characteristic spores which contain amoeboid sporoplasms and are usually provided with filaments that can be extruded from the spore. Two major types may be recognized. The myxosporeans, which grow in the body cavities and tissues of their hosts, have a spore which is a complex multicellular structure, normally constructed of two or more valves enclosing one or more sporoplasms and several polar filaments coiled within polar capsules. The microsporeans, which are normally intracellular parasites, have a spore which is a unicellular structure containing a single sporoplasm and often a single filament which is not coiled within a polar capsule.[121] Few species of either of these classes are well-known, and accounts of the two groups will take the form of a description of a representative example, followed by comments on features of other members of the group.

The presence in the life cycle of Myxosporea of a multicellular stage in which some cells of the body are purely somatic and have no reproductive future, has been used as an argument in favour of the removal of this group from the Protozoa to the Metazoa. The absence of a flagellate stage and the presence of an amoeboid sporoplasm are other features which isolate these from other Protozoa. The presence of extrusible polar filaments comparable with those in nematocysts does not seem a good enough reason to relate them with coelenterates, since dinoflagellates possess similar organelles (p. 131). Perhaps further study of the ultrastructure and life cycle of these organisms will throw some light on their relationships with other animals.

CLASS MYXOSPOREA[145, 165, 201]

Myxobolus pfeifferi, a member of the order Myxosporida, causes a 'boil' disease of cyprinid fish, as a result of which vast numbers of spores escape from burst cysts in the surface tissues of the fish. At one end of each spore are two polar capsules, each containing a coiled polar filament, and in the remaining space between the two shell valves is a single mass of sporoplasm that is initially binucleate, but later appears to have only a single nucleus (Fig. 10.1). When a spore is swallowed by a suitable host fish, the polar filaments are everted and presumably serve for anchorage in the gut of the host. The sporoplasm emerges as a small amoeba, passes through the gut wall and migrates, possibly in the blood vessels, to the muscles and connective tissues of the body wall. Here the parasite grows and its nucleus divides to form a syncytial mass which may break up and spread. The host tissue around the parasite becomes modified to form a thick envelope and a cyst is formed that may be easily visible to the naked eye. Certain nuclei of the syncytium become surrounded by dense cytoplasm and a cell membrane to form cells that differentiate as sporonts. The sporont grows and divides twice to produce two large cells and two small cells; the latter form a coat around the larger cells, each of which will differentiate to form a complete spore. The nucleus of each large

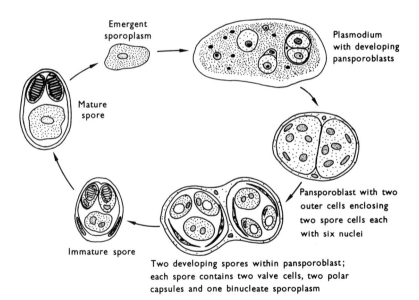

Fig. 10.1 The life cycle of *Myxobolus pfeifferi* (see text).

cell divides forming six nuclei; two of the resulting cells form the valves, two form the pair of polar capsules and the two remaining nuclei remain in the central sporoplasm (Fig. 10.1). After maturation the spore remains dormant, but, following its release by rupture of the cyst, it may be swallowed by a new host and commence a new cycle of growth.

Other members of the order Myxosporida are also parasites of cold-blooded vertebrates, principally fish. They are best identified by spore characters, since they differ in the number and shape of shell valves and in the number and arrangement of polar capsules (Fig. 10.2). Some species develop in body cavities of the host, normally the gall bladder or urinary

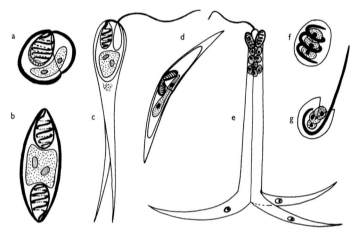

Fig. 10.2 The spores of some myxosporeans; a, *Unicapsulina muscularis* (6 μm) from halibut: b, *Myxidium lieberkuhni* (20 μm long) from pike: c, *Henneguya* sp. (50 μm long) from freshwater fish, with one polar capsule everted: d, *Ceratomyxa arcuata* (25 μm long) from gall bladders of marine fish: e, *Triactinomyxon ignotum* (150 μm long) from gut epithelium of the worm *Tubifex*: f, g, *Helicosporidium parasiticum* from dipteran larvae, showing complete spore (5 μm) and a hatching spore with filamentous cell and three sporoplasms.

bladder, while other species grow in the tissues, in muscle, cartilage, liver, etc. In all cases the parasite seems to be extracellular and feeds on fluids; it is amoeboid with short pseudopodia and is said not to be phagotrophic. Some of these species are of economic importance since they parasitize fish that are commercially valuable; they may reduce the growth of the fish, spoil the flesh of the fish or kill them, and sometimes the diseases may become epidemic.

Another order in this class is the Actinomyxida, members of which have spores with three valves, three polar capsules and often numerous sporoplasms; they are parasites of the body cavity and gut epithelium of oligochaete worms and sipunculids, e.g. *Triactinomyxon ignotum* in *Tubifex* (Fig. 10.2e). A third order, the Helicosporida, contains a single species parasitic in arthropods. The spores of this species contain three uninucleate sporoplasms surrounded by an elongate cell formed into a single coiled filament, and are enclosed in a single membrane (Fig. 10.2f, g).

CLASS MICROSPOREA[145, 201, 245]

Two species of the genus *Nosema* (Order Microsporida) cause epidemic diseases of economic importance. Spores of *Nosema apis* may be found

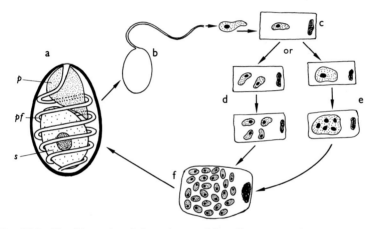

Fig. 10.3 The life cycle of the microsporidian *Nosema*; a, the mature spore with sporoplasm(*s*), polar filament (*pf*) and polaroplast (*p*): b, the sporoplasm emerges from the everted polar filament: c, the sporoplasm enters a host cell and grows, and may either divide repeatedly d or may become multinucleate before division e: finally f the host cell contains many sporonts, each of which develops into a complete spore.

in the cells of the midgut wall and malpighian tubules of adult honey bees, in which this parasite causes the destructive Nosema disease. The spores are about 5 μm long and 3 μm wide and have a thickened membrane, at one pole of which is attached an eversible tubular filament that is coiled around the single, uninucleate sporoplasm in the intact spore (Fig. 10.3). Following the ingestion of a spore by the host, the polar filament is everted, possibly by expansion of the polaroplast, and the amoeboid sporoplasm emerges through the tubular filament and enters the cells of the gut epithelium. Within the cells the parasite grows and undergoes asexual reproduction and sporogony, so that the host cell becomes distended and hypertrophied. Each sporont produced in the host

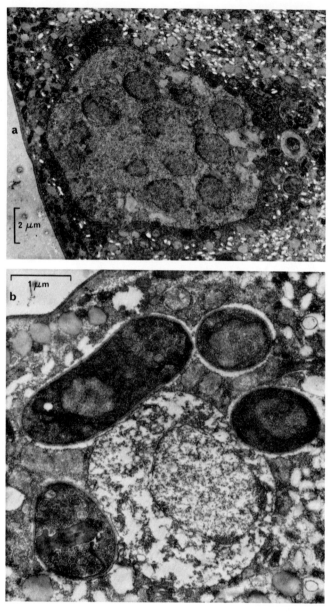

Fig. 10.4 Electron micrographs of sections of a microsporidian parasitic in the sporozoan *Selenidium*. a, A pansporoblast stage with a number of nuclei in a common cytoplasm; b, several spores surrounding a uninucleate sporoblast. Micrographs by A. G. H. Dorey.

cell develops into a single spore. These spores may be shed into the gut and passed out with the faeces to spread the infection to new hosts. Another important species of the same genus, *Nosema bombycis*, is responsible for the pebrine disease of silkworms, in which tissue cells of any type may be infected at any stage of growth, so that even eggs may be infected and the larvae die without spinning the silk coccoons.

Microsporidians are intracellular parasites, going through their entire development within a host cell. It appears from electron micrographs that the host cell does not produce a vacuole membrane around the parasite, as is usual in Sporozoa (p. 238), so that only a single unit membrane separates parasite cytoplasm from host cytoplasm. Electron micrographs

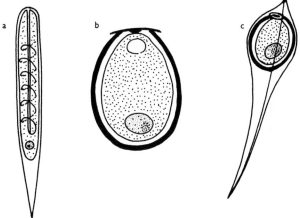

Fig. 10.5 The spores of some microsporidians; a, *Mrazekia caudata* (20 μm) from the worm *Tubifex*: b, *Haplosporidium limnodrili* (10 μm) from the gut of the worm *Limnodrilus*: c, *Urosporidium fuliginosum* (10 μm, excluding projection) from the worm *Syllis*.

of two stages in the development of a microsporidian parasitic in the sporozoan *Selenidium* are shown in Fig. 10.4.

Members of the order Microsporida generally invade the muscles, intestinal epithelium, lymphocytes and adipose tissue of invertebrates, especially insects, but almost all groups have some microsporidian parasites. Thus, species of *Nosema* have been recorded from Protozoa,[276] helminths, annelids, arthropods, fish and mammals, including man. In the single definite human case, spores were found in the cerebrospinal fluid and urine, and in diseased rodents the same parasite, *Nosema cuniculi*, has been found in the brain and occasionally more widely distributed in lymphoid tissues. A common disease of sticklebacks (*Gasterosteus*) and some other freshwater fish is caused by species of *Glugea* which cause the formation of white cysts, up to 5 mm across, in the skin or muscles of the fish; other species of this genus are found in amphibians and reptiles. In spite of their widespread occurrence the microsporidians are still poorly known, and certainly many species remain to be studied and described; spore characters are important in classification (e.g. compare Fig. 10.3a with Fig. 10.5a).

Spores containing a single amoeboid sporoplasm but no polar filament characterize the order Haplosporida (Fig. 10.5b, c), which is now normally classed as a second order of the class Microsporea. These parasites are found in invertebrates and lower vertebrates, but are poorly known. The sporoplasm develops into a plasmodium within which the spores differentiate. Although these spores do not possess an eversible filament, they are sometimes provided with rigid spines formed from the spore coat, and may have some form of operculum. *Ichthyosporidium* is found in fish and *Haplosporidium* in annelids and molluscs.

II

Ecology of Protozoa

THE ROLES PLAYED BY PROTOZOA IN ECOSYSTEMS

The trophic structure of a generalized ecosystem is illustrated in Fig. 11.1; the solar energy captured by the photo-autotrophs provides the chemical energy essential for the life of organisms in the other three categories. Flagellate organisms are extremely important as primary producers of organic matter in aquatic habitats; in particular, dinoflagellates

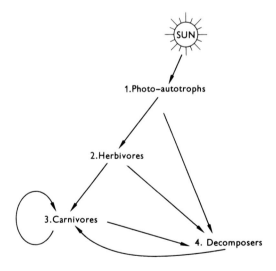

Fig. 11.1 Component trophic levels in a generalized ecosystem.

among the larger forms and Haptophyceae and Chrysophyceae among the smaller flagellates may form very large populations. Autotrophic flagellates are also of importance as symbionts within the cells of a variety of organisms.

Many smaller plants are eaten by ciliate, amoeboid and flagellate Protozoa. The food of such Protozoa includes diatoms and coccoid algae as well as flagellates; some amoebae and ciliates are able to ingest filamentous algae, especially the more easily fragmented blue–green filaments. Protozoa are rather rarely found as parasites in plants.

Predatory forms are again found in all three groups of free-living Protozoa. In addition to eating other Protozoa they are also found to ingest metazoan larvae and such smaller Metazoa as rotifers. Parasitic Protozoa are found in animals of all groups.

A role is played in the decomposition of organic matter by a diversity of Protozoa. Fragments of detritus are engulfed by various ciliates and amoebae as well as by a few flagellates, and a wide range of Protozoa of all three groups are able to take up dissolved organic matter and make use of it as a source of energy and raw materials. These organisms are competing with bacteria, fungi and other organisms for the food content of dead organic matter. Many Protozoa feed on bacteria, and so channel back some of the energy content of the decomposer group into the higher trophic levels of the food chain, as the Protozoa may be eaten by other animals; organisms that consume bacteria are of particular importance in the soil and in polluted waters. These Protozoa are also thought to play a significant part in nutrient cycles.

THE INFLUENCE OF ECOLOGICAL FACTORS ON THE LIVES OF PROTOZOA[67, 142, 190]

Physical and chemical factors

Protozoa with the ecological roles outlined above occur in almost any body of water, salt or fresh, permanent or temporary, as well as in damp soil, damp moss, snow and within the bodies of other animals and plants. These habitats provide conditions within the range of ecological tolerance of Protozoa, the more stringent habitats requiring more particular specializations of the protozoan inhabitants. Ecological factors of particular importance in the life of various Protozoa are water, temperature, oxygen, pH and salinity; if these factors are within favourable limits for a species of protozoon, then its occurrence and abundance will depend upon the availability of suitable food and the extent of predation.

Active trophic Protozoa need *water* because an area of unprotected cell membrane is necessary for purposes of feeding, and such active

Protozoa are therefore absolutely limited to damp environments. Protozoa are most abundant in aquatic habitats, but are also characteristic inhabitants of soils and polar regions where free water is only present for short periods; Protozoa from these habitats are able to escape shortage of water by encystment, which will be considered in more detail later (p. 260). Protozoa inhabiting soils, moss, temporary pools or snow have the ability to excyst rapidly on the return of favourable conditions, as one may see in the occurrence of *Chlamydomonas nivalis* ('red snow') in the thin films of water in part-melted snow. Among organisms included with the Protozoa, those that are most tolerant of dry conditions are probably the slime moulds, found on damp, rotten wood or decaying vegetation, but they also have a means of resisting desiccation by the formation of dry sclerotia as well as reproductive spores.

Dormant stages of Protozoa, including cysts, spores and sclerotia, survive extreme **temperatures**; for example, the cysts of the soil ciliate *Colpoda* have been found to germinate after immersion in liquid air for seven days or heating to 100°C for three hours. The lower temperature limit for the active life of Protozoa is provided by the freezing point of the surrounding water, and many Protozoa will grow and reproduce at temperatures down to almost 0°C in fresh waters and about −2°C in sea water. The optimum temperature for the life of a strain of protozoon seems to be at least partly a matter of acclimation; the same species of organism may be found in cold streams at below 5°C and in hot springs at 40–50°C, but 'cold strains' generally have a lower optimum temperature for growth and division than 'hot strains'. Some species, by contrast, are characteristic of more restricted temperature ranges; for example, the ciliate *Trochilia palustris* has been found to be characteristic of colder waters, and burst when it was allowed to warm up in the laboratory, while another ciliate *Halteria grandinella* was characteristic of warmer waters. Similarly, some species of foraminiferans and marine dinoflagellates are found in polar seas, while other species are found in tropical waters.

The protozoan fauna of hot springs is of some interest, for freshwater Protozoa are commonly found in laboratory studies to be unable to survive temperatures above 30–35°C. Several species of testaceous amoebae have been found in hot springs at 40–45°C, the ciliates *Nassula elegans* and *Cyclidium glaucoma* up to 51°C and the amoeba *Hyalodiscus* up to 54°C. There have been reports of Protozoa in hot springs of Ischia in the Bay of Naples at temperatures of 53–64·7°C. Comparable high temperatures may occur in shallow pools in direct sunlight, and a variety of ciliates has been found under such conditions at temperatures of 50–52°C.

The optimum temperature for the growth of a protozoon in laboratory culture may vary with the food provided, e.g. *Euglena gracilis* grown in the dark in a casein proteose medium without acetate had a maximal division rate at 10°C, while identical cultures grown with acetate had a maximal division rate at 23°C. In an experimental study of the growth and division of *Tetrahymena pyriformis* it was found that although growth

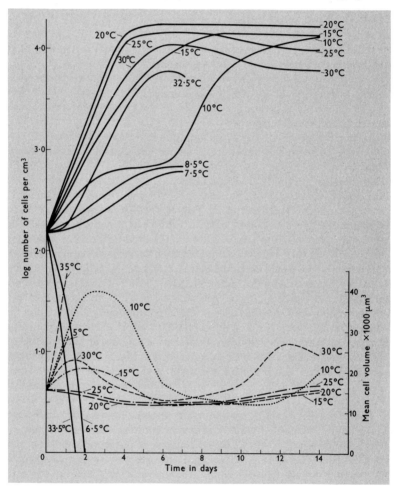

Fig. 11.2 The effect of temperature on the population density and cell size in cultures of *Tetrahymena pyriformis*. The continuous lines indicate the changes in population density at different temperatures following the introduction of ciliates from a culture maintained at 20°C into tubes of proteose peptone. The broken lines indicate the changes in mean cell volume at different temperatures. Data by J. Bullock (unpublished).

was possible over the whole experimental range from 5°C to 35°C, division (i.e. population increase) only occurred at temperatures between 7·5°C and 32·5°C (Fig. 11.2). The volume of the ciliates remained low at about 20°C (the normal culture temperature), but high volumes of 2–3 times the normal were reached at extreme temperatures. E. Zeuthen and his colleagues have used temperature changes to control the time of division in *Tetrahymena* and thereby obtained cultures in which the stages of the cell cycle were synchronized (see p. 98). Similar division-synchrony may be achieved with the flagellate *Astasia* using heat treatment and with photosynthetic flagellates using alternate periods of dark and light.[224]

An important feature of the environment that varies with temperature is the amount of **oxygen** dissolved in the water; the amount of oxygen that will dissolve in water decreases from about 14 p.p.m. at 0°C to about 7·5 p.p.m. at 30°C. Since the metabolic rate of organisms, and hence their utilization of oxygen, also tends to increase with rise in temperature, oxygen is more likely to limit the abundance of aerobic organisms in warmer waters. The oxygen content of fresh waters varies seasonally and diurnally, depending on the extent of photosynthetic and respiratory activities of the organisms present. Few free-living Protozoa seem to be capable of living in the complete absence of oxygen, but many tolerate very low levels of oxygen, provided other substances common in anoxic situations, such as H_2S and CO_2, are not present at high concentrations. A few species, including *Trepomonas agilis*, *Caenomorpha medusula*, *Pelomyxa palustris* and *Saprodinium putrinum*, are characteristically found in anaerobic freshwater habitats, and *Trepomonas* soon dies if the water is aerated. A number of other forms such as *Spirostomum ambiguum*, *Stentor coeruleus*, *Amoeba proteus*, *Actinosphaerium*, *Difflugia*, *Peranema*, and many more, are often found in regions of low oxygen content, and are at least capable of enduring oxygen lack if not of thriving without oxygen. While some gut parasites may be facultative anaerobes like many of these freshwater forms, it is believed that the flagellates in the gut of termites are obligate anaerobes, and it is likely that the ciliates in the rumen contents of cattle and sheep live in a medium that is almost continuously anoxic. It is clear that some Protozoa make use of anaerobic pathways of respiratory metabolism, and it is likely that many more need very little oxygen although they cannot rely exclusively on anaerobic respiration.

The **carbon dioxide** content of water tends to vary inversely with the dissolved oxygen. Most Protozoa are fairly tolerant of low CO_2 concentrations, but high concentrations are toxic to many species, particularly those from situations that are normally well oxygenated. Thus, forms from oligosaprobic lakes, such as *Codonella*, *Ceratium hirudinella* and *Synura uvella* were found to be very sensitive to CO_2, while *Paramecium putrinum* and *Polytoma uvella*, characteristic of polysaprobic conditions, were tolerant of high CO_2 concentrations. The CO_2 content of water is vital in that it provides the carbon source of autotrophic organisms, but it is also important because it combines with water to form carbonic acid and tends to lower the pH of the water. Because of the diurnal fluctuations of photosynthetic activity it is commonly observed that the pH of water rich in algae rises considerably in the day and falls at night. In correlation with this pH change, it is interesting that the pH-sensitive species *Spirostomum ambiguum* was found to be free-swimming by night but to retreat into the layer of decaying leaves by day when the pH rose.

The *pH* of the environment may vary for other reasons. Certain waters may become very acid because of the release of humic acids in the absence of buffering salts which would otherwise permit bacterial action; such waters contain a characteristic protozoan fauna. Even more extreme acid conditions are found in some cases where industrial effluents or mine drainage enter streams. Records of Protozoa from such habitats indicates great tolerance by some species, extending to a pH of 1·8 in the case of *Euglena mutabilis*, *Chlamydomonas* sp., *Urotricha* sp. and *Oxytricha* sp. In acid soils testaceous amoebae may be the commonest organisms.

Protozoa belonging to some groups are found only in the sea and those of others only in fresh water, but there are few species that are able to live in both habitats. The difference in *salinity* between sea and fresh water involves a substantial change in osmotic pressure and in the concentrations of various ions, and it is therefore a considerable feat for a protozoon to survive the transition from fresh water to sea water or the reverse. Such a transition generally involves an immediate shrinkage or swelling, followed presumably by a leakage of salts in or out and a subsequent return towards normal body size; changes in the activity of the contractile vacuole are normally involved (see p. 23). While it has been found to be possible to transfer *Cyclidium glaucoma* directly from fresh water to sea water, and to grow the freshwater *Amoeba lacerata* in salinities of up to 44‰, a majority of forms probably have a much more restricted tolerance, e.g. *Paramecium caudatum* would live at a salinity of about 15‰ but not above, while *Cryptomonas ovata* var. *palustris* would not even tolerate a rise in salinity to 0·3‰. Very high salinities occur in certain situations in salt lakes and salt marshes, and in such habitats the ciliate *Fabrea salina* was found at a salinity of 200‰, *Prorodon utahensis* and *Uroleptus packi* are ciliates that were found in water with a salt content of 23% and the flagellate *Dunaliella salina* has been recorded from waters with a salt content above 250 g l^{-1} in Lake Elton.

Pigmented Protozoa are most sensitive to *light*, since they absorb a greater part of its energy. Light provides the energy source for photosynthetic organisms, and many of these show phototactic migrations, both in water and more solid media, as exemplified by the swimming of *Volvox* and *Euglena* towards light of moderate intensity, and the diurnal migrations of marine dinoflagellates in the plankton, of *Amphidinium* in sand and of *Euglena deses* in mud. Interestingly enough, a phototactic response is also shown by some green ciliates—ciliates containing symbiotic zoochlorellae. Exceedingly strong light is damaging; many Protozoa may respond to high light intensities by negative phototaxis, as in *Volvox* and some species of *Euglena*, while in some other species of *Euglena* (*E. sanguinea* and *E. haematodes*) a protective red pigment is found to migrate to the body surface in bright light and to 'migrate inwards in weak light, and a number of other flagellates that live in very light places tend to carry red pigmentation, e.g. *Chlamydomonas nivalis* in 'red snow'.

Food and its effect on population size

The distribution of various Protozoa in nature corresponds with the availability of essential nutrients or of the favoured type of food, provided that the physico-chemical characteristics of the habitat are within tolerable limits. The general nutritional needs of diverse Protozoa have been discussed in Chapter 3. The needs of autotrophs and of heterotrophs which ingest dead organic matter are likely to be easily satisfied in a diversity of habitats, but predatory heterotrophs like *Didinium nasutum* (eating *Paramecium*) and *Actinobolina radians* (eating *Halteria*) are very specific in their food requirements, and only occur where their prey may be found in plenty. A more varied diet is found in many ciliates, amoebae and flagellates which feed on bacteria of a wide range of types, although some Protozoa have been found to favour one collection of bacterial species and other Protozoa favour another collection of bacterial species; certain pigmented bacteria such as *Chromobacterium violaceum* are toxic.[51] Some of the ciliates and larger amoebae may feed on a wide variety of algae and on other Protozoa as well as on organic detritus and on some small multicellular animals such as rotifers and larvae of various sorts. A protozoon with a more diversified diet may be able to exploit a greater range of ecological niches.

Many laboratory studies have been made of the growth of Protozoa in culture. A typical **growth curve** for a population of *Tetrahymena pyriformis* growing on a proteose–peptone medium is shown in Fig. II.3. There is an initial lag phase (I) of slow multiplication, which is absent if the inoculum of ciliates is taken from an actively growing culture; this is followed by an exponential growth phase (II), a phase of decreasing growth rate (III), a stationary phase of maximum population (IV), and a slow death phase (V). The slope of the curve in the exponential growth phase and/or the maximum population attained may vary with such factors as temperature and with the availability of any limiting factor in the diet;

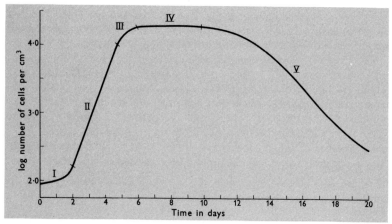

Fig. 11.3 Phases in the growth of a population of *Tetrahymena pyriformis* in culture.

the energy source may not be limiting, and in such cases the stationary phase may be prolonged, lasting perhaps for a month or more in *Tetrahymena*.

The size of the maximum population attained under given culture conditions varies with the quantity of food available. The data shown in Fig. 11.4 indicate that not only are there more *Tetrahymena* present at

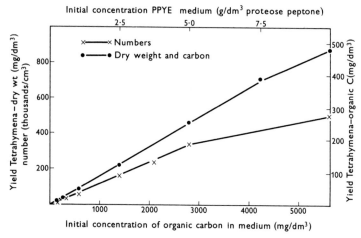

Fig. 11.4 The effect of food concentration, expressed as the concentration of proteose peptone/yeast extract medium and as organic carbon, on the numbers and yield (dry weight and organic carbon) of *Tetrahymena pyriformis* in culture. Data by Curds and Cockburn,[46] with permission of H.M.S.O.

higher concentrations of organic food, but that the individual ciliates grow larger when more food is available. As a result of such studies C. R. Curds and A. Cockburn[46] found that about 9% of the organic carbon in the food was converted into organic carbon in ciliates. It appears that about $\frac{3}{4}$ of the organic material in this proteose–peptone/yeast extract medium could not be utilized by the ciliates, and that about $\frac{2}{3}$ of the carbon assimilated was oxidized to carbon dioxide in respiration.

When Curds and Cockburn[46] grew *Tetrahymena* monoxenically on various concentrations of the bacterium *Klebsiella aerogenes*, they found a relationship between food concentration and the population of *Tetrahymena* similar to that obtained in axenic culture. In this case, however, the dry weight **growth efficiency** (increase in dry weight of *Tetrahymena*/dry weight of bacteria eaten) was about 41%, which is much higher than the value obtained when the ciliates were feeding on the peptone medium, and probably reflects the fact that bacteria are a natural food of *Tetrahymena*, so that a smaller proportion of the assimilated food is required for respiration to produce energy for chemical interconversions. At high concentrations of bacteria the feeding mechanism of *Tetrahymena* becomes saturated at a maximum level. If the concentration of bacteria was kept constant and the population density of *Tetrahymena* was raised, the feeding rate of the ciliates decreased at higher ciliate concentrations, presumably because of interference between the feeding activities of adjacent ciliates. The interrelationships between the bacterial concentration, the ciliate population and the feeding rate of the ciliates is shown in Fig. 11.5; it is evident that the food consumption of individual ciliates is related to the ratio of bacterial concentration to ciliate concentration. Curds and Cockburn[49] also found that smaller *Tetrahymena* cells grew more quickly than larger ones at a particular bacterial concentration and that ciliates of a particular size grew more quickly as the bacterial concentration was increased until a maximum growth rate for that size of ciliate was reached (Fig. 11.6); this diagram was derived from measurements made on organisms grown in a two-stage continuous culture system in which a growth efficiency of up to 54% could be obtained.

Feeding rates in situations where the predatory protozoon is a 'hunter' rather than a 'grazer' show similar features modified by the particular relationship between predator and prey. For example, *Didinium nasutum* hunts for and ingests *Paramecium* (Fig. 3.5, p. 65), but the proboscis structures used in catching the prey must be reformed after each meal,[289] and as a result there is a minimum interval between meals of about 2 hours at 20°C. Each meal taken by *Didinium* is a large one, for the relative volumes of *Didinium*:*Paramecium* range between about 1:1 for a large, well-fed *Didinium* and 1:5 for a small, starved *Didinium*.[24] If the prey are abundant *Didinium* may eat a *Paramecium* every two hours, but if

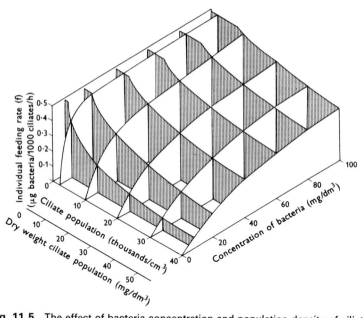

Fig. 11.5 The effect of bacteria concentration and population density of ciliates on the individual feeding rate of *Tetrahymena pyriformis* on the bacterium *Klebsiella aerogenes*. Data by Curds and Cockburn,[46] with permission of H.M.S.O.

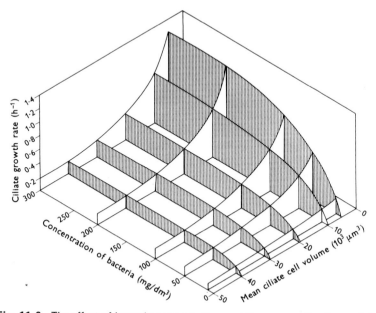

Fig. 11.6 The effect of bacteria concentration and the mean cell volume of the ciliates on the specific growth rate of *Tetrahymena pyriformis*. Data by Curds and Cockburn,[49] with permission of H.M.S.O.

the prey are scarce the rate of feeding is less because of the time taken in hunting after the proboscis is reformed. The regulation of feeding rate by the availability of prey at low densities of *Paramecium* and by the feeding interval characteristic of the predator at high densities of prey is indicated in Fig. 11.7, where it is evident that above a ratio of prey: predator of about 20:1 the *Didinium* are feeding at a maximum rate, each catching a *Paramecium* every 2 hours. The absolute density of the prey and predator is of course also important, and it is seen that at the lowest density of *Didinium* tested the maximum feeding rate of *Didinium* is not

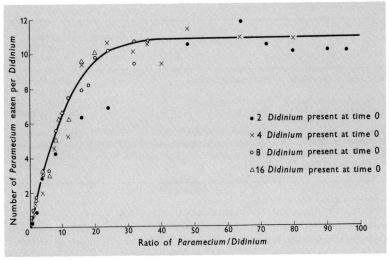

Fig. 11.7 The number of *Paramecium* eaten in a 20 hour period by each *Didinium* present at different initial ratios of *Paramecium*: *Didinium* for four initial population densities of *Didinium*. In calculating the feeding rate, the number of *Paramecium* consumed is divided by the average number of *Didinium* present during the 20 hour period to take account of the increase in number of *Didinium*; the *Paramecium* had been starved, and control experiments showed that very few of them would divide during the 20 hour period. Data by P. Brobyn.[24]

attained until the prey:predator ratio is considerably above 20:1. The absolute density of prey and predator also have an important effect on the dry weight growth efficiency (see p. 255); this varies from about 8% when 2 didinia and 48 paramecia were placed in a volume of 13·8 cm³ (timed over 20 hours) to about 25% when 2 didinia and 48 paramecia were placed in a volume of 5·2 cm³ (20 hours). If the *Didinium* can find no prey they encyst, but this form of cyst does not resist drying, although it provides for survival in times of food shortage and is viable in the hydrated state for some years.

It is interesting that the growth efficiencies of these Protozoa are in general rather higher than the values commonly quoted for metazoan organisms. It is true that the values are not strictly comparable because it is more meaningful to express growth efficiency in terms of energy content (calorific value) rather than dry weight, but unfortunately data of this type are not yet available for Protozoa. Some examples may be seen in Table 11.1. From this table it appears that the high figures avail-

Table 11.1 · The growth efficiency of a number of Protozoa and other animals growing under specified conditions.

Animal	Food	Growth efficiency (%)	Basis of estimate
Entodinium caudatum	*Escherichia coli*	50	Carbon (growth/ingestion)
Tetrahymena pyriformis	*Klebsiella aerogenes*	41·4	Dry weight
Tetrahymena pyriformis	*Proteose peptone*	36·5	Carbon (growth/ingestion)
Acanthamoeba sp	*Saccharomyces cerevisiae*	37	Dry weight
Didinium nasutum	*Paramecium caudatum*	8–25	Dry weight
Beef cattle (overall)	Grass	4·1	Calorific value
Beef cattle (young)	Grass	~35	Calorific value

able for Protozoa are taken from situations where the conditions were suitable for rapid growth, and such figures are indeed comparable with growth efficiencies for young growing Metazoa. In more generalized situations, which probably include stationary phases of little population growth interspersed with periods of exponential growth, one might expect to obtain values within the range of those quoted for the whole life cycle of a metazoan organism. It is nevertheless interesting that very high growth efficiencies may be achieved under such conditions as are maintained for example in the aeration tank of an activated sludge sewage plant (p. 270).

In natural situations it is likely that there will be **competition** between Protozoa for the available resources. G. F. Gause[84] studied competition between *Paramecium aurelia* and *Paramecium caudatum* in many different types of situation. In one case, when the two ciliates were grown separately on bacterial food under a given set of conditions both grew to fairly high populations, but when they were grown together under the same conditions *P. aurelia* attained higher populations and biomass than *P. caudatum*. In another set of experiments the two ciliates were grown under another set of conditions with yeast for food, and in this case *P. caudatum* attained a higher biomass than *P. aurelia* when they were grown together

(Table 11.2). Under one set of conditions the larger *P. caudatum* grows more quickly and suppresses the growth of *P. aurelia*, but under the other set of conditions *P. aurelia*, which is apparently more resistant to toxic products, multiplies more quickly and eventually eliminates *P. caudatum* completely. Such experiments with standardized conditions and limited volumes are clearly artificial in comparison with a small part of a pond or stream where emigration, immigration and a greater diversity of food

Table 11.2 The maximum populations achieved by *Paramecium aurelia* and *P. caudatum* grown separately or together on two different types of food. The 'relative biomass' of these two ciliates is 0·39:1. Information from Gause.[84]

Food	Culture of the two *Paramecium* species	*P. aurelia*		*P. caudatum*		Ratio $\frac{P.\ aurelia}{P.\ caudatum}$	
		Numbers	Relative biomass	Numbers	Relative biomass	Numbers	Relative biomass
Pseudomonas	Separately	192	75	135	135	1·42	0·55
pyocyanea	Together	146	57	44	44	3·32	1·30
Saccharomyces	Separately	212	83	145	145	1·46	0·57
exiguus	Together	152	59	80	80	1·90	0·74

and physico-chemical conditions exist, but they do give some impression of the complexity of interactions which must occur in nature—a small change in some feature of the environment may profoundly affect the balance of competition between two species. It is of some interest, then, that over longer periods of time various habitats tend to contain a similar range of species in rather similar proportions, and one must suppose that this tendency towards balance is maintained by the overall stability of environmental conditions in the absence of severe pollution.

Relationship to substrata

Although there are many planktonic Protozoa among the radiolarians, flagellates, tintinnid ciliates and foraminiferans, the majority of Protozoa live in communities in association with some environmental structure, moving over its surface or between components of the substratum. This is illustrated by the small, temporary communities of organisms (micro-biocoenoses), involving algae or filamentous bacteria and Protozoa, in which the algae or bacteria provide both a structural substrate and a source of food for an entire food web of Protozoa; several of these associations will be described below (p. 265), and in each case it is interesting that the density of Protozoa is very much higher within the community than it is in the surrounding water.[196]

An association which may be closely comparable with these involves psammophilic forms dwelling in the interstices between the grains of some marine sands,[74] where the Protozoa probably collect bacteria from the surfaces of the sand particles. The Protozoa involved in these associations become attached to surfaces more or less firmly by pseudopods, stalks or 'thigmotactic cilia', and many of them use these temporary attachments as an anchorage while they graze over the surface or create feeding currents in the surrounding water. Similar associations occur between Protozoa and the surfaces of some larger animals and plants.

Survival through adverse conditions

Many Protozoa secrete resistant walls within which they remain dormant through periods of adverse conditions. In some cases this cyst wall is a relatively permeable layer and the encysted organism cannot survive

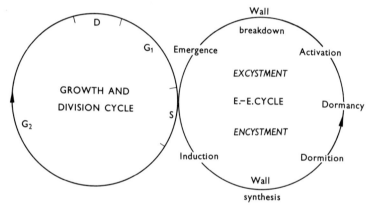

Fig. 11.8 The relationship between the growth-division cycle and the encystment-excystment cycle of *Acanthamoeba*. Modified from a figure by Neff and Neff.[184]

drying out, e.g. the cyst of *Didinium*. The cysts of soil Protozoa, of some parasites and of Protozoa from temporary bodies of water are able to survive drought and other severe conditions; the walls of such cysts are highly resistant and rather impermeable.

The soil amoeba *Acanthamoeba* encysts in response to starvation, desiccation, overcrowding, accumulation of metabolic wastes or raised temperature. The life cycle and encystment of these amoebae has been studied experimentally by R. J. and R. H. Neff[184] (Fig. 11.8). Encystment was promoted by transfer from a nutrient growth medium to an inorganic salt medium containing calcium and magnesium ions, in which it is thought that failure to complete the S phase of the growth/division cycle led to the induction of encystment. It is likely that all adverse conditions causing encystment result in an interference with DNA synthesis, and hence block the cell cycle at the S phase. In the experiments the induction process was found to lead to a rounding up of the cell and a massive breakdown of cell components, including many mitochondria and storage granules. It was followed by two waves of wall synthesis, the first produced an outer phospho-protein layer and the second an inner cellulose layer; these synthetic reactions involved specific RNA synthesis and the formation of new enzymes. All detectable metabolism ceased within a few days after the completion of the cyst wall, there was no longer any cytoplasmic movement within the cyst, and the organism entered a period of dormancy in which it could survive for years. In the dormant state such cysts can endure high and low temperatures, drying, low and high pressures and absence of oxygen. Comparable detailed information on the induction of excystment of *Acanthamoeba* is not available, but, following an initial stage of activation that is dependent upon RNA synthesis and protein synthesis, there is some digestion of the cellulose wall and a progressive increase in activity of the organism, and within 12 to 30 hours the amoeba emerges. A number of studies of excystment of the soil ciliate *Colpoda* indicate that soluble substances present in hay infusions, principally organic acids and sugars, provide the necessary stimulus for induction of excystment in this organism, so that it is presumed that the resistant cyst wall is permeable to such substances.[280] In the amoebo-flagellate *Naegleria* the response of the amoeba to changed conditions includes encystment with drying and the development of flagella with dilution of the medium (p.164).

FEATURES OF THE LIFE OF PROTOZOA IN VARIOUS HABITATS

It was estimated by V. A. Dogiel[67] that out of 20 000 known species of Protozoa about 5000 are parasites. The Sporozoa and Cnidospora are entirely parasitic, and there are parasitic representatives in several groups of flagellates, of ciliates and of amoebae. About two thirds of the remaining 15 000 free-living species are marine forms, including such large groups as the Radiolaria (∼6000 spp.) and Foraminiferida (∼1000 spp.) and the majority of the dinoflagellates (∼1000 spp.) and tintinnid ciliates (∼800 spp.). The Heliozoa, testaceous amoebae, Euglenophyceae and Chlorophyceae are predominantly freshwater organisms.

Protozoa from marine habitats[67]

The most extensive marine habitats are those of the open ocean. The

vast majority of **oceanic Protozoa** are planktonic forms found within
100 m or so of the surface of the sea, in the region where abundant food
is available because of photosynthetic activity in the illuminated zone.
In deeper waters, even at extreme depths, planktonic Protozoa occur,
but these abyssal forms are almost exclusively Radiolaria (Phaeodaria).
Planktonic Protozoa are classed as mesoplankton, organisms with a diameter exceeding 1 mm, microplankton, organisms with a diameter between 50 μm and 1 mm, or nannoplankton, organisms with a diameter
of less than 50 μm. Some of the largest radiolarians, especially colonial
species, come within the mesoplankton, and many of the smaller flagellates, principally from the Haptophyceae, Silicoflagellida and Chrysophyceae are classed as nannoplankton. The plants of the nannoplankton
are frequently responsible for the bulk of the primary production, and
it has been estimated that in some waters there may be 1000 or more
times as much chlorophyll in the nannoplankton as in larger organisms;
the majority of these small forms are flagellates, and the other plants
will include dinoflagellates and diatoms. In some waters the dinoflagellates
symbiotic in radiolarians contribute a significant proportion of the primary
production; in tropical waters the Acantharia are especially abundant,
but are less numerous in colder waters. The radiolarians and a few species
of foraminiferans are the characteristic planktonic amoebae, with long
spines and often oil droplets to aid flotation and long slender pseudopods

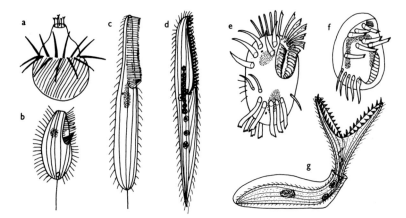

Fig. 11.9 Some ciliate Protozoa common in shallow benthic marine habitats. a,
Mesodinium pulex 25 μm (Gymnostomatida); b, *Uronema marina* 40 μm
(Hymenostomatida); c, *Cohnilembus verminus* 100 μm (Hymenostomatida); d,
Gruberia lanceolata 500 μm (Heterotrichida); e, *Diophrys appendiculata* 80 μm
(Hypotrichida); f, *Aspidisca steini* 30 μm Hypotrichida); g, *Folliculina aculeata*
350 μm (Heterotrichida) and further examples in Fig. 11.10

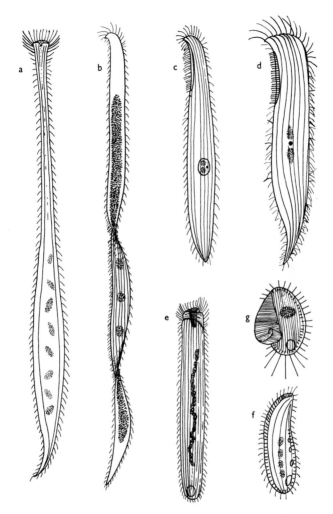

Fig. 11.10 Some ciliates from marine sand; a to f belong to the Gymnostomatida and g to the Hymenostomatida. a, *Trachelocerca phoenicopterus* 1 mm; b, *Centrophorella (Kentrophoros) lanceolata* 500 μm, ribbon-shaped, with a dense mat of sulphur bacteria on the dorsal surface; c, *Geleia fossata* 350 μm; d, *Remanella rugosa* 200 μm, seen from the dorsal surface but showing the ciliary meridians of the ventral surface; e, *Chaenia gigas* 500 μm; f, *Lionotus* sp. 60 μm; g, *Pleuronema coronatum* 80 μm; and some of the examples shown in Fig. 11.9 are also found in marine sands.

for the capture of food. The tintinnid ciliates and some of the dino-flagellates are other important predators among the microplankton.

A wider range of protozoan species is found where there are substrata for attached forms. These are occasionally present in oceanic regions where there are numbers of epiphytic and epizoic Protozoa such as peritrich and chonotrich ciliates attached to crustaceans. In **coastal regions** there are abundant surfaces, inanimate as well as living, upon which Protozoa may settle. The majority of species of foraminiferans are benthic species, many of them from shallow waters, and there are characteristic species of flagellates and ciliates which occur in close association with or attached to various forms of substratum;[23] these include dinoflagellates like *Amphidinium* and *Oxyrrhis*, and ciliates like vorticellid peritrichs, hypotrichs like *Euplotes* and *Diophrys*, gymnostomes like *Trachelocerca*, suctorians like *Ephelota*, heterotrichs like *Condylostoma* and the tubicolous *Folliculina* (see also Fig. 11.9). The interstices between the grains of certain marine sands are inhabited by some specialized ciliates[74] (Fig. 11.10); these psammophilic forms have been studied by J. Dragesco,[68, 69] who found that they were either small species, or extremely flattened and often elongate ciliates, showing marked thigmotactic attachments to the sand grains over which they crept while collecting bacteria and organic detritus.

Freshwater Protozoa[190, 216]

Freshwater habitats are somewhat more diverse and include puddles and temporary bodies of water as well as lakes and rivers. The larger lakes have a true open-water plankton, including among the autotrophic forms such dinoflagellates as *Ceratium* and *Peridinium*, such chlorophyceans as *Pandorina* and *Volvox* and smaller chrysophycean forms; in freshwater phytoplankton these flagellates may be outnumbered by blue–green algae, desmids, diatoms and other coccoid forms. The availability of nutrient salts required for the growth of plants influences the abundance of phyto-plankton in lakes. Waters with few salts are referred to as oligotrophic, and the rather sparse phytoplankton is dominated by characteristic diatoms and desmids and such chrysophycean flagellates as *Dinobryon*, *Mallomonas* and *Synura*; eutrophic waters with a higher content of nutrient salts support a rich and diverse flora, often dominated by blue–green algae, and with a different range of diatoms and desmids, and often in-cluding dinoflagellates, although *Ceratium* and some species of *Peridinium* may also be found in oligotrophic waters. Among the zooplankton of lakes there are a few freshwater representatives of the tintinnid ciliates, e.g. *Codonella*, and some other ciliates, colourless flagellates and helio-zoans are commonly present, but generally it is found that as in the sea the coastal (neritic) regions have a richer fauna and flora. Smaller ponds provide similar habitats to those found around the edges of larger fresh-water lakes. In streams and rivers the range of protozoan species is very similar to that characteristic of ponds and lakes, except for the greater proportion of attached forms and a smaller proportion of free-floating forms, especially in places where the current is strongest. However, even in quite turbulent streams there are still some motile Protozoa, particularly ciliates whose thigmotactic ciliature enables them to maintain close contact with suitable surfaces.

The species of organisms found in all of these freshwater habitats vary according to the amount of *organic matter* present, as well as according to such features as the oxygen content, calcium, nitrate and phosphate concentrations and pH. Water with a high organic content is described as polysaprobic, and waters with medium and low organic contents are referred to as mesosaprobic and oligosaprobic respectively, while extremely pure waters are described as katharobic. It is possible to encounter any of these categories of organic content in waters that are classed as eutrophic on the basis of mineral salt levels, but oligotrophic waters will generally have rather low levels of dead organic matter. The largest numbers of protozoan organisms occur in polysaprobic conditions, but normally rather few species are represented, principally bacteria-consuming forms tolerant of very low oxygen concentrations, such as the flagellates *Oikomonas mutabilis* and *Bodo* spp., the ciliates *Paramecium putrinum* and *Vorticella microstoma*, forms like *Polytoma uvella* that depend upon dissolved organic matter, and a few photosynthetic forms like *Euglena*. A richer, more diverse range of Protozoa is found in mesosaprobic conditions, including some pigmented flagellates like *Chlamydomonas* and *Cryptomonas* as well as such colourless flagellates as *Chilomonas*, *Anthophysa* and *Peranema*, ciliates like *Colpidium*, *Carchesium*, *Vorticella* spp., *Stentor*, *Euplotes* and *Aspidisca*, the heliozoans *Actinophrys* and *Actinosphaerium* and such amoebae as the testacean *Arcella*. Photosynthetic forms tend to predominate in oligo-saprobic conditions, including such flagellates as *Dinobryon*, *Gonium*, *Volvox* and *Ceratium*, accompanied by a rather sparse fauna of ciliates and amoebae that are mostly herbivorous or detritus feeders.

As in the case of neritic marine Protozoa, the majority of species tend to be associated with surfaces, and in particular with the surfaces of fila-mentous plants. A number of workers have recognized quite characteristic associations or *communities* (microbiocoenoses) involving Protozoa in such situations (p. 259). For example, E. Fauré-Fremiet[75] described the part played by ciliates in the life of a mat of filaments of the sulphur bacterium *Beggiatoa*. At first the mat of filaments was fairly clean and there were a few *Colpidium* eating free-swimming bacteria around the *Beggiatoa*. Increase in the number of these ciliates was followed by the arrival of predatory ciliates like the gymnostome *Lionotus* and the hymenostome *Leucophrys* (= *Tetrahymena vorax?*), and a reduction in the numbers of *Colpidium* led to an increase in the numbers of *Glaucoma* eating *Beggiatoa* and *Chilodonella* eating associated blue–green algae or filamentous bacteria. A rapid multiplication of these last two ciliates led to the breakdown of the mat of *Beggiatoa* and the production of detritus which forms the food of such ciliates as *Paramecium* and the gymnostome *Coleps*. At all stages the number of ciliates closely associated with the mat of filaments was much higher than that in the surrounding water or on the adjacent mud.

The close association between Protozoa, even many of the more active swimming ciliates, with filamentous growths of algae and bacteria, had been emphasized in an earlier study by L. E. R. Picken.[196] In waters of lower organic content the filamentous growths of blue–green algae, particularly *Oscillatoria*, formed the basis of the community of organisms, and in waters of high organic content the organisms were associated with growths of 'sewage fungus' (a complex association in which the filamentous bacterium *Sphaerotilus* is usually dominant). In both communities there were Protozoa feeding on bacteria and on diatoms, but the numbers of different Protozoa varied with the relative quantities of these two major forms of food, diatoms being more abundant in cleaner water and bacteria being abundant in polluted water. Both communities included forms consuming detritus and carnivorous Protozoa as well as some omnivorous species; some relationships between the organisms involved are shown in Fig. 11.11. It is interesting that the Protozoa maintained their close

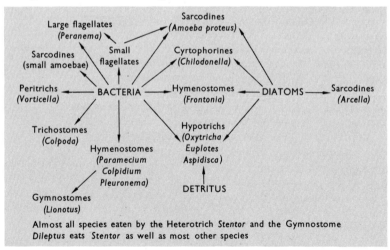

Almost all species eaten by the Heterotrich *Stentor* and the Gymnostome *Dileptus* eats *Stentor* as well as most other species

Fig. 11.11   The major protozoan components of a freshwater food web where the sources of food are bacteria, diatoms and detritus. Data from Picken.[196]

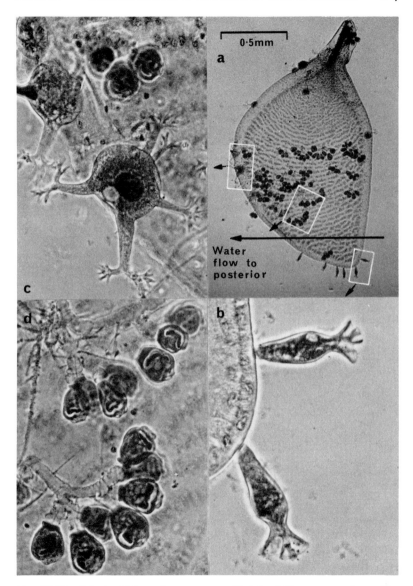

Fig. 11.12 Epizoic ciliate protozoans on a gill plate of *Gammarus pulex*. a, The whole gill plate stained to show up the Protozoa, and labelled to indicate the orientation of the gill plate in the water current created by limb movements of the crustacean; b, *Spirochona gemmipara* (Chonotrichida); c, *Dendrocometes paradoxus* (Suctorida); d, *Epistylis* sp. (Peritrichida).

contact with the filamentous surfaces even when the diatoms were also common on the surrounding mud, so that it seems as if the specific association may depend upon mechanical and chemical characteristics of the substratum, the filaments of algae and bacteria providing a preferred micro-environment.

An association of species on brackish-water, intertidal mud-flats studied by M. Webb,[285] involved a comparable community in which the basic algal substrate was provided by a film of diatoms on the surface of the mud, and the Protozoa included forms feeding on bacteria, diatoms and detritus as well as carnivorous forms (Table 11.3).

Table 11.3 The diet of some of the ciliates found in a community on estuarine mud of the river Dee. Information from M. Webb.[285]

Ciliate Protozoa		Food			
Genera	Order	Diatoms	Bacteria	Detritus	Protozoa
Loxophyllum, Trachelocerca, Lacrymaria	Gymnostomatida (Rh)				√
Chilodonella	Gymnostomatida (Cy)	√			
Frontonia	Hymenostomatida	√			
Cohnilembus, Pleuronema	Hymenostomatida		√		
Vorticella	Peritrichida		√		
Condylostoma	Heterotrichida	√			
Spirostomum	Heterotrichida		√		
Blepharisma	Heterotrichida		√	√	
Strombidium	Oligotrichida	√			
Stylonychia	Hypotrichida		√		
Aspidisca	Hypotrichida		√	√	
Diophrys	Hypotrichida	√	√		
Euplotes, Uronychia, Opisthotricha	Hypotrichida	√	√	√	

Relationships between Protozoa and **animal surfaces** are also found, but are more restricted and probably have a different basis. Many of the Protozoa involved are sessile forms e.g. in the epizoic associations of the peritrichs *Vorticella* or *Epistylis*, the chonotrich *Spirochona* and the suctorian *Dendrocometes* on the gill plates of *Gammarus*. Here the water currents provided by the crustacean may provide an abundant supply of food for the ciliates, probably carrying different sorts of food to the different parts of the gill plate normally occupied by the three Protozoa (Fig. 11.12). Motile epizoic forms include such hypotrichs as *Kerona*, found on *Hydra*, and mobiline peritrichs like *Trichodina*, *Urceolaria* and *Cyclochaeta*, which are common on the respiratory organs of fish, on polychaetes and on many other animals; in some cases these ciliates have to be regarded as parasites because they harm the animal providing the support, but in general it appears that they are harmless bacteriophagous forms.

Protozoa, pollution and sewage treatment

Most streams and rivers show variations in the content of organic matter along their length because of the inflow of organically polluted water from tributaries.[120] Such organic materials are broken down by the organisms living in the stream as the water flows along, and at a distance below the site of pollution the organic materials will have disappeared and all that remains is an enriched water containing more mineral salts than were present in the original stream water. Close to the site of pollution there will be intense bacterial activity associated with extensive utilization of oxygen in the breakdown of the organic compounds; if the pollution is extensive, the oxygen content may be reduced to zero and the water may contain hydrogen sulphide and large amounts of nitrogenous compounds, especially ammonia. In such anaerobic conditions only a few eucaryote organisms occur, including such flagellates as *Bodo* and *Oikomonas*, along with the abundant swimming bacteria and carpets of sewage fungus. As the organic matter is broken down and the consumption of oxygen is reduced, aeration and the reappearance of some green plants cause the oxygen content to rise again, and the general character of the water reverts from the polysaprobic condition towards the mesosaprobic and ultimately to the oligosaprobic condition. Protozoa that eat bacteria will be most abundant in more polluted conditions, but with the appearance of plants, herbivorous ciliates and amoebae become common.[75a] Laboratory studies on artificial systems with added organic matter show similar chemical changes and biological successions.[21a]

Severe pollution that results in extensive deoxygenation does great damage to the life of the stream and should not be permitted. Such damage is avoided by the treatment of polluted waters to reduce the content of organic compounds and suspended solids before the water enters the stream; the various forms of sewage treatment achieve a reduction in the amount of organic material by bacterial action in the same manner as is normal in a stream, but the process is accelerated and is isolated from the stream.

Three processes of **sewage treatment** are in common use; Protozoa occur abundantly and are probably of importance in all three processes.[45, 174] Small-scale treatment may be carried out in a septic tank (Imhoff tank). This is a large chamber into which the heavily polluted water flows, and within which there are abundant bacteria and large amounts of organic material to oxidize, with the result that the medium only contains oxygen in a shallow surface zone. The main part of the organic breakdown takes place under anaerobic conditions, with the use of hydrogen acceptors other than oxygen, so that much hydrogen sulphide and ammonia is usually produced; the effluent from such a tank is therefore normally only partially purified. The deeper levels of the tank contain abundant anaerobic bacteria and a number of Protozoa that are facultative or obligate anaerobes, particularly species tolerant of H_2S.[190] The ciliates *Metopus es*, *Trimyema compressa* and *Saprodinium putrinium* and the flagellate *Trepomonas agilis* appear to be obligate anaerobes, and other common forms in such tanks include the amoebae *Euglypha alveolata* and *Vahlkampfia guttula* and the flagellates *Cercobodo caudata*, *Pleuromonas jaculans* and *Hexamita inflata*.

Where sewage treatment is carried out on a larger scale, the liquor that remains after removal of solids that can be settled or skimmed off is subjected to processes in which aerobic oxidation of the organic compounds is promoted, either by allowing the liquor to percolate through beds of stone separated by continuous air spaces, or by passing the liquor into chambers (activated sludge tanks) where it is intensively aerated. In the former system, the percolation filters contain stones (larger ones below and smaller ones down to about 4 cm in diameter at the surface) arranged in a bed almost 2 m deep in such a way that air can circulate readily through spaces between the stones. A slow trickle of the sewage liquor runs down over the stones, and a bacterial film quickly appears on the stones and develops into a complex association of aerobic bacteria, fungi, blue–green algae and Protozoa.[47] Organotrophic forms promote aerobic organic breakdown and bacteria, algae, detritus and Protozoa provide food for the Protozoa that are present in the water film which covers the stones. A range of Protozoa that might be found in such a situation is indicated in Fig. 11.13. Such larger animals as nematode and oligochaete worms and insect larvae are abundant, and are important in that they burrow through the bacterial film and break it up, so that fragments are washed away and the air passages are kept open.

In the **activated sludge process** the bacteria that perform the main part of the aerobic oxidation of organic matter are principally swimming forms, but some of them aggregate together by a flocculation process that is enhanced by the presence of ciliated Protozoa, possibly because of mucoid secretions of the ciliates.[50] The formation of these flocs removes many bacteria from suspension and provides substrates for the attachment of stalked and thigmokinetic ciliates. The most abundant ciliates are bacterial feeders, but some species also eat detritus or are carnivores. Since the oxygen content is kept high by bubbling with air or by rotating paddles, all the species involved are aerobic. After the liquor has spent some hours in the tank it passes to a settlement tank where the flocculated material is removed from the effluent; some of the settled material is

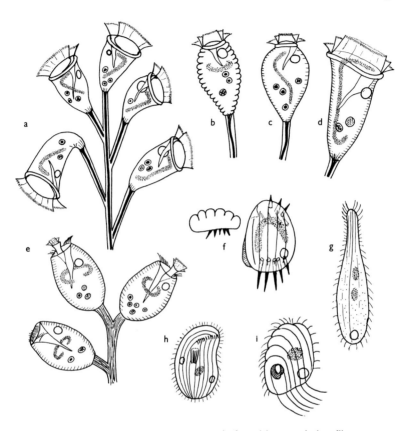

Fig. 11.13 Some ciliate protozoa commonly found in percolation filter sewage treatment systems. a, *Carchesium polypinum*, cell body 125 μm long; b, *Vorticella striata*, cell body 40 μm long: c, *V. microstoma*, cell body 50 μm long; d, *V. convallaria*, cell body 75 μm long; e, *Opercularia microdiscum*, cell body 80 μm long (a–e all Peritrichida); f, *Aspidisca costata* 30 μm (Hypotrichida), seen from the posterior end and from dorsal surface; g, *Trachelophyllum pusillum* 40 μm (Gymnostomatida); h, *Chilodonella uncinata* 60 μm (Gymnostomatida); i, *Cinetochilum margaritaceum* 30 μm (Hymenostomatida).

returned to the activated sludge tanks to maintain a high population of microorganisms.

C. R. Curds and his colleagues have recently undertaken a number of studies on Protozoa of sewage treatment plants. In a survey of 52 percolating filter systems in Britain,[47] ciliates formed the most abundant group, although the testate amoeba *Arcella vulgaris* was commonly present and the flagellate *Peranema trichophorum* was frequent; the more important

Table 11.4   The ciliate Protozoa recorded most commonly in a survey of sewage treatment plants in Britain.   In each case the species are listed in order of their frequency of occurrence in plants using that type of treatment, and the percentage of cases in which the ciliates were found in very large numbers is shown alongside. Information from Curds and Cockburn.[47]

Percolating filters		Activated-sludge plants	
Species	% with high numbers	Species	% with high numbers
Chilodonella uncinata	4	*Vorticella microstoma* *	10
Vorticella convallaria *	10	*Aspidisca costata* †	35
Opercularia microdiscum *	44	*Trachelophyllum pusillum*	15
Carchesium polypinum *	15	*Vorticella convallaria* *	19
Opercularia coarctata *	2	*Opercularia coarctata* *	12
Aspidisca costata †	—	*Vorticella alba* *	11
Cinetochilum margaritaceum	—	*Euplotes moebiusi* †	5
Vorticella striata *	2	*Vorticella striata* *	2
Trachelophyllum pusillum	—	*Vorticella fromenteli* *	4
Opercularia phryganeae *	4	*Carchesium polypinum* *	8

* Peritrich ciliates.   † Hypotrich ciliates.

ciliates found are listed in Table 11.4. Of 56 activated sludge plants studied in Britain,[47] three contained no ciliates, but in the others ciliates were the dominant Protozoa; *Arcella vulgaris* and *Peranema trichophorum* were found in these also, and a somewhat different range of ciliates (Table 11.4). The role of these Protozoa in sewage treatment is more easily studied in activated sludge plants because the inner parts of a functional percolating filter are less accessible. In studies of small-scale experimental activated sludge systems, Curds found that following the setting up of a plant the protozoan fauna developed in a clear sequence.[44] At first only flagellates (*Oikomonas, Peranema, Bodo*) were present, but after several days a range of 'free-swimming' ciliates like *Paramecium* appeared and became the dominant forms after about two weeks. These ciliates were replaced after about 4 weeks by crawling ciliates, especially species of the hypotrich *Aspidisca* and stalked peritrich ciliates, mainly *Vorticella* spp. Finally, in the mature system *Aspidisca* and *Vorticella* appeared to be competing for dominance. While the majority of the ciliates were feeding on bacteria, some, e.g. *Lionotus*, were carnivores and some, e.g. *Trachelophyllum*, were thought to ingest sludge flocs. Experimental activated sludge plants maintained free of ciliates produced turbid effluents with a high B.O.D. (biological demand for oxygen), high quantity of suspended solids and large numbers of viable bacteria; after the addition of ciliates to these plants, the clarity of the effluent improved and the

B.O.D., suspended solids and viable bacteria all decreased dramatically, high numbers of ciliates being correlated with low numbers of bacteria in the effluent. The ciliates are assumed to effect this improvement primarily by feeding on bacteria, but the enhancement of flocculation and the promotion of the micro-circulation of liquor by ciliary activity may also play a part. In the survey of activated sludge plants mentioned above, it proved possible to predict the quality of effluent produced by a plant from a knowledge of the range of ciliate species present;[48] in general it is found that a good quality effluent results where the species composition corresponds to that of a river showing oligosaprobic or cleaner mesosaprobic characteristics. Computer simulation techniques have been successfully applied to protozoan and bacterial populations in activated-sludge systems.[45a]

Soil Protozoa[250]

At one time it was thought that there were no active Protozoa in the soil, and that such Protozoa as were found there were encysted and originated from freshwater or marine sources. Early in this century however, it was established that there was a considerable fauna of active feeding Protozoa living in the water films around soil particles, where they were presumed to be feeding on soil bacteria; estimates indicated that there are between a few thousand and half a million Protozoa in each gramme of moist soil, along with perhaps 10^9 bacteria. From a study of soils from many parts of the world H. Sandon[221] concluded that all soils contained Protozoa, principally in the surface layers, and that similar species occurred in soils at all latitudes. About 20 of the 250 protozoan species he found in the soil were only present in that habitat, while the majority of species were also present in organically polluted waters. In general the abundance of flagellates, ciliates and amoebae could be correlated with features associated with good bacterial growth, while the abundance of testate amoebae was correlated with the amount of organic matter, irrespective of bacterial numbers; many of the Protozoa could live where the oxygen concentration was very low or even zero. Many studies in more recent years have confirmed these conclusions and extended the lists of species of Protozoa found in and specific to soils. Several reports give the impression that there is a correlation between the numbers of Protozoa present and the vegetative productivity of the soil, and there are reports that nitrogen fixation in soils is enhanced in the presence of Protozoa, although the means by which this enhancement is achieved are not clear. V. F. Nikoljuk,[187] who has contributed much to knowledge in this area, has also found that Protozoa secrete such plant activators as 3-indylolacetic acid (heteroauxin). It is common experience that the populations of Protozoa in the rhizosphere areas immediately around plant roots contain the same species as the surrounding soil in considerably larger numbers; generally bacteria are also more abundant in the rhizo-sphere.[53]

The common soil Protozoa include the flagellates *Oikomonas, Cercomonas* and *Heteromita*, several species of the ciliate genus *Colpoda*, the amoebae *Hartmannella, Acanthamoeba* and *Naegleria* and species of the acrasian myxomycete *Dictyostelium*. Many testate amoebae reach their greatest abundance in soils, and may be the dominant Protozoa in acid soils where there are few bacteria; such genera as *Centropyxis, Nebela, Trinema* and *Euglypha* are commonly represented. It is likely that under most conditions the flagellates and small amoebae are more important than the larger amoebae or ciliates, e.g. a 1 g sample of soil from an English wheatfield contained 1500 amoebae, 32 000 flagellates and 20 ciliates, but in some soils records of thousands of ciliates and up to 70 000 testate amoebae in each gramme of soil indicate that these larger Protozoa may sometimes be very important in the biological economy of the soil.

The majority of these soil Protozoa encyst and therefore have a means of surviving dry conditions. Such cysts also provide a means of distribution of these Protozoa; it has been found that there are on average 2 protozoan cysts/m³ of air, and it is thought that these minute dry cysts may be carried for some distance by air currents. A more common means of dissemination of soil species as well as freshwater Protozoa is probably by the movements of animals, which may carry wet or dry mud, slime, etc. from place to place; a number of Protozoa have been recovered from mud attached to the plumage and legs of birds and even from slime taken from the body of migratory water beetles.

SYMBIOTIC RELATIONSHIPS INVOLVING PROTOZOA

Epibiotic species

Many Protozoa live epibiotically on the surfaces of other animals and plants, frequently using these organisms only as a support; stalked and mobiline peritrich ciliates provide many examples of this mode of life in both marine and freshwater habitats and there are even epizoic peritrichs living in water films on the gills of terrestrial woodlice. A number of such epizoic forms have made the transition to parasitism, changing their mode of life to a greater or lesser extent to derive food and often other needs such as shelter and dispersal from their hosts. A range of forms may be seen in the ciliate order Thigmotrichida (p. 204), members of which live among the gills in the mantle cavity of bivalve molluscs; the less specialized forms feed on detritus and use the gills only as a supporting substratum, while the more advanced forms show a progressive development of a sucker or tentacle system and are thought to obtain their nourishment more directly from the host. Many species of the mobiline peritrich *Trichodina* appear to live a harmless epizoic existence, but numerous other species are parasites on the gills of their hosts, e.g. fish, and some have become endoparasites within the urinary bladder of fish. The majority of ectoparasitic Protozoa are ciliates, although there are a few dino-flagellates and amoebae following this mode of life.

Endoparasitic Protozoa[158]

Protozoa are seldom found as parasites in terrestrial plants. Perhaps the best-known examples are the flagellates (*Phytomonas*) found in the latex channels of *Euphorbia* spp. and some other plants, and which are assumed to be transmitted by plant-sucking bugs. There have also been reports of amoebae inhabiting similar situations in euphorbias, figs and lettuce plants.

Multicellular animals provide a number of suitable habitats for parasitic Protozoa, the different locations demanding rather different means of transmission from host to host. The least specialization is required of parasites inhabiting the alimentary canal of animals, for means of entry with the food and exit with the faeces are already provided, and the parasite has only to solve the problems of avoiding desiccation before being eaten by another host and avoiding being killed and digested by the digestive juices of the host. Many gut-dwelling Protozoa eat bacteria just as their relatives do in freshwater, and by secreting protective cyst walls they survive exposure to the air and exposure to the digestive juices of the host which eats them. As with many parasites, the chances of successful transmission to a new host are enhanced by the production of enormous numbers of infective individuals—in this case cysts. It is likely that more than half of the parasitic Protozoa live in such situations, inhabiting a specific part of the alimentary canal. Many of these forms attach themselves to the gut wall by means of suckers, hooks or tentacular appendages; they often grow to considerable size and many have very complex structures. While an abundance of bacteria in the gut provides food for many Protozoa, many more feed by absorbing the food digested by the host, and a third category derive nourishment from the gut epithelia of the host, either by engulfing the cells or by extracting nutrients through appendages that penetrate between the cells; those of the third category are clearly most harmful to the host.

A number of parasites that gain entry to the body through the mouth and alimentary canal proceed to invade the tissues and body cavities of the host. Thus many coccidians invade the epithelia of the alimentary tract and associated organs and release infective cysts into the gut following growth and reproduction. Some gregarines and cnidosporans migrate to the body cavity and surrounding organs of the host, where they grow and produce cysts or spores. These pass out through coelomoducts, and are liberated when the host moults or lays eggs, or in some cases are only released at the death of the host. Occasionally the parasites enter the eggs of the host and are therefore passed directly to a host of the next generation (e.g. *Babesia*). Some tissue parasites gain entry to the new host through the skin, especially through less protected areas, such as gills.

Protozoa inhabiting the blood spaces of vertebrates differ from these other forms in adaptive ways related to their mode of life. These parasites generally have a simple structure and are either very small if they live within corpuscles, or have a small diameter if they live free in the plasma, because of the need to pass through blood capillaries. Since the blood systems are enclosed, the parasites need a special means of entry and exit; this is frequently provided by blood-sucking vectors which transmit the protozoon from one host to another, arthropods being vectors for terrestrial vertebrates and leeches for aquatic vertebrates. The parasite normally grows and reproduces in the vector also, and is a true parasite in both of its hosts. Because of the direct transmission between vertebrate and vector, these Protozoa do not form resistant cyst walls. It seems likely that the relationship between the Protozoa and their two hosts may have arisen differently in the flagellates and in the Sporozoa. The flagellates (trypanosomids) were probably originally intestinal parasites of invertebrates, as are *Leptomonas* and *Crithidia* today, and some of these became secondarily adapted to live part of their life in vertebrate blood systems. On the contrary, it is thought that the sporozoan blood parasites were probably derived from intestinal parasites of the vertebrate hosts which became parasites in the vertebrate blood system, perhaps following an intermediate stage similar to that of the eimeriid *Schellackia*. In this parasite the sporozoites escape into the blood from the site of zygote formation in intestinal epithelial cells of a lizard and are subsequently taken up by a blood-sucking mite, the cycle being completed by a lizard eating the mite. It is important that in these less elaborate life cycles, including that of *Lankesterella*, which shows a prolonged phase in the vertebrate blood-stream, the protozoans show no changes in the invertebrate host, and it is only in more complex cycles of such genera as *Haemoproteus* and *Plasmodium* that multiplicative phases occur in the arthropod.

The ***immunology*** of parasitic infections by Protozoa has been the subject of intensive study, particularly with respect to diseases caused by malaria parasites and trypanosomes.[252] The hosts possess a diversity of mechanisms that protect the body against the entry of foreign organisms or which act to remove foreign organisms that gain entry into the body. Non-specific immunity is provided by such cellular mechanisms as those which isolate the parasite by encapsulation, or digest it following phagocytosis by macrophages. Specific immunity to a particular species of parasite is somewhat more delayed and is conferred in vertebrates by the production of antibodies in response to the presence of antigenic substances of that parasite. This is a complex subject, and the range of effectiveness of the immune response of the host can be seen by the fact that some infections result in the death of the host, some in the death of the parasite and some in a chronic infection in which host and parasite live in some sort of equilibrium. Chemotherapeutic treatments are employed to tip the balance in favour of the host.

In total there are probably about 5000 parasitic Protozoa, including all of the Sporozoa and Cnidospora, about one quarter of the flagellates, one sixth of ciliates and one eightieth of all amoebae. It is interesting that a majority of the protozoan parasites so far described have been found in terrestrial hosts, parasites belonging to the Kinetoplastida, Metamonadida, Entodiniomorphida, Coccidia and Haemosporidia being particularly numerous in terrestrial hosts, while Cnidospora and Opalinida are more common in aquatic hosts.

Symbiotic associations of mutual benefit

In a number of cases the Protozoa that live within other animals are thought to be beneficial to the host rather than harmful. Such mutualistic associations include the ciliate populations in the alimentary canals of herbivorous mammals and the flagellate fauna of the intestine of termites.[67, 117, 138] In the case of cattle and sheep the rumen (anterior stomach) contains in each cm^3 of gut contents up to half a million ciliate Protozoa of many (30–50) species, mostly belonging to the order Entodiniomorphida (p. 213), but with large numbers of trichostome ciliates in the genera *Isotricha* and *Dasytricha*. These ciliates are obligate anaerobes that are transmitted from host to host in saliva; the grass from the rumen is regurgitated for chewing, and ciliates may survive for a short time on dropped grass or in drinking water. Although the ciliates thrive in the alkaline or mildly acid conditions of the rumen, they are quickly killed in the acid conditions of the posterior stomach (omasum). Some of these Protozoa ingest and break down plant fragments and others take up sugars from the gut contents. It has been estimated that about $\frac{1}{5}$ of the protein needs of cattle are met by the digestion of ciliates which are killed as they pass down the alimentary canal, and that the absorption of the products of cellulolytic activity of the ciliates provides about $\frac{1}{5}$ of the energy needs of the host. The relationship of these activities of Protozoa to bacterial activity in the rumen is not yet clear. The large intestine of horses contains similar numbers of ciliate Protozoa, but it is not known what contribution these may make to the efficiency of digestion within the host.

Flagellate Protozoa are of even greater importance to wood-eating termites and roaches and it has been estimated that between one third and one seventh of the weight of a nymph of *Zootermopsis* is provided by Protozoa in the hind gut. The flagellates are obligate anaerobes which digest the cellulose of the wood eaten by the host insect and benefit the host because some of the products of digestion released by the Protozoa are absorbed by the insect as its principal source of food. In the absence of flagellates the insect starves to death, although it continues to eat wood; the infection of termites with flagellates is normally ensured by the habit of termites eating the fresh faeces of other termites. The hypermastigote and polymastigote flagellates (p. 147) are found only in these insect hosts, and are as dependent upon the insect as the insect is upon them. The ecological importance of these flagellates is considerable, for in some regions of the world termites replace earthworms as agents of humus breakdown.

Relationships in which other organisms live on or in Protozoa[10]

Protozoa are themselves subject to parasitic infection by other Protozoa. These include small amoebae parasitic within Opalinids, in foraminiferans and in other amoebae, and parasitic flagellates, such as leptomonads in *Paramecium* and dinoflagellates in radiolarians, in tintinnid ciliates and even in the nucleus or cytoplasm of other dinoflagellates. Most of the ciliates which parasitize Protozoa are suctorians living on or in other ciliates, although there have also been reports of thigmotrichs living ectoparasitically on peritrichs and suctorians. A number of protozoan parasites are themselves the hosts of other parasitic protozoans; microsporidians of the family Metchnikovellidae are frequently found in gregarine sporozoans (p. 244), and a species of *Nosema* has been found as a parasite of a myxosporidian from fish.

Protozoa may also act as hosts for organisms of other types. Bacteria have often been found attached to the outer surfaces of Protozoa, in particular on flagellates from termites and sand-dwelling ciliates. H. Kirby described the presence of spirochaetes and of rod-shaped and fusiformis-like bacteria arranged in various specific patterns on the surfaces of termite flagellates. Some flagellates form surface structures to which these epibionts attach. In *Mixotricha paradoxa* spirochaetes and bacteria are arranged in a precise pattern in relation to projecting 'brackets' on the surface of the flagellate; it has been shown that the swimming of this flagellate depends upon the undulations of these spirochaetes, which move in a coordinated fashion with clear metachronal waves. The rod-shaped bacteria that cover the dorsal surface of psammophilic ciliates of the genus

Centrophorella (Fig. 11.10b) appear to be sulphur bacteria, for they often contain sulphur granules; there seems to be some specificity in the relationship between these bacteria and these ciliates, for the bacteria could not be found on surrounding sand grains.

Many Protozoa contain bacteria, frequently gathered around or concentrated within the nucleus; such bacteria are especially frequent in flagellates from termites, and in ciliates from the stomach of ruminant mammals, and they have also been reported from amoebae, opalinids and free-living flagellates and ciliates. Little is known of the significance of the presence of these bacteria in most cases, but two types of example have received some attention. Rod-shaped bacteria present in the cytoplasm of the flagellate *Crithidia oncopelti* confer upon this flagellate the ability to grow on a lysine-free medium containing only four amino acids and lacking haemin; *Crithidia* lacking the bacteria required lysine, haemin and 12 amino acids. Symbiotic organisms that are almost certainly bacteria have been found living within the cytoplasm of *Paramecium aurelia*[129] and have been given such names as kappa, mu, pi, lambda, etc.; lambda 'particles' carry appendages that look like bacterial flagella, and several types of particle have cell walls of a bacterial type. It was found many years ago that paramecia from certain stocks consistently killed paramecia from other stocks when the two were mixed, and on account of this interaction the former were called killers and the latter sensitives. Killer paramecia were found to contain kappa particles, but the persistence of kappa in the cytoplasm was dependent upon the presence of a dominant macronuclear gene. Only kappa particles that contained refractile inclusions (tightly-coiled ribbons associated with virus-like granules) were able to kill, but other kappa particles could develop these inclusions. The other forms of particle act in a comparable manner. Paramecia that possess mu kill sensitives at conjugation (they are mate killers), while pi may be an inactive (mutant?) kappa, and paramecia containing lambda produce large ($\sim$5–10 μm) lethal granules which are released into the medium. It is assumed that these symbiotic organisms derive nutrients from the paramecia, but a killer stock has an advantage over a sensitive stock in ecological competition.

Relationships between Protozoa and fungi appear to be one-sided in favour of the fungi, for in all cases it appears that the fungi either parasitize Protozoa or are predaceous on them—though these categories of 'attack' are not sharply separated. Pigmented organisms that are attacked by chytrid fungi include flagellates of various groups—Chrysophyceae, Dinophyceae, Euglenophyceae, Chlorophyceae and probably others. These fungi may enter the flagellate cell and develop within it, or may send root-like hyphae into the flagellate and withdraw the contents; in either case the flagellate is usually killed, and the fungus grows and produces flagellate zoospores that provide a means of dispersal. In addition to attacking these free-living aquatic flagellates, organisms that are assumed to be chytrids (mostly attributed to the genera *Sphaerita* and *Nucleophaga*) have been found within the cells of flagellates from termites and other hosts, in parasitic amoebae, in rumen ciliates from mammals and in gregarines from arthropods, some within the nucleus and others in the cytoplasm. Soil amoebae, both naked and testate, are attacked by fungi of other groups, the Zoopagales and Hyphomycetes. These fungi either enter the cell and grow within the cytoplasm, perhaps following ingestion within food vacuoles, or entrap the amoeba with adhesive hyphae before penetrating it with hyphal roots. An infection of *Paramecium bursaria* with a yeast has been reported.

Relationships involving photosynthetic organisms[10, 243]

Algal cells that are found within or upon Protozoa seem to be living with one another on more balanced terms. Blue–green algae, dinoflagellates and green algae (Chlorophyceae) all occur within Protozoa, and dinoflagellates and chlorophyceans also live symbiotically within the cells of multicellular animals, principally coelenterates and molluscs. Blue–green algal symbionts are referred to as cyanellae, and the associations of Protozoa with cyanellae have been termed syncyanosen. Cyanellae found externally on the chrysophycean *Oikomonas syncyanotica* appear to contribute to the nutrition of the flagellate, for motility of the colourless flagellate bearing cyanellae would occur in the absence of free oxygen, but was light dependent. Other syncyanosen involve cyanellae that live within amoebae and flagellates, e.g. the association of the blue–green alga *Cyanocyta korschikoffiana* within the cryptomonad *Cyanophora paradoxa*.

Dinoflagellates living in such mutually beneficial associations are termed zooxanthellae and are found in various marine organisms including Protozoa. Although they occur in large numbers in radiolarians and occasionally in foraminiferans and ciliates, little is known about the physiological relationship between the zooxanthellae and the host Protozoa. Investigations on the zooxanthellae (usually regarded as *Symbiodinium* sp.) of anemones, corals and clams show that the algal cells excrete such products of photosynthesis as glycerol, alanine, glucose and organic acids, but that

---

Fig. 11.14 Electron micrograph of a section through the surface region of the flatworm *Convoluta roscoffensis* showing the elongate cells of the intracellular algal symbiont *Platymonas*. The algal cells contain extensive plastid systems and a pyrenoid, and in some cases (at arrow) a flagellar basal body may be seen. Micrograph by A. E. Dorey (unpublished).

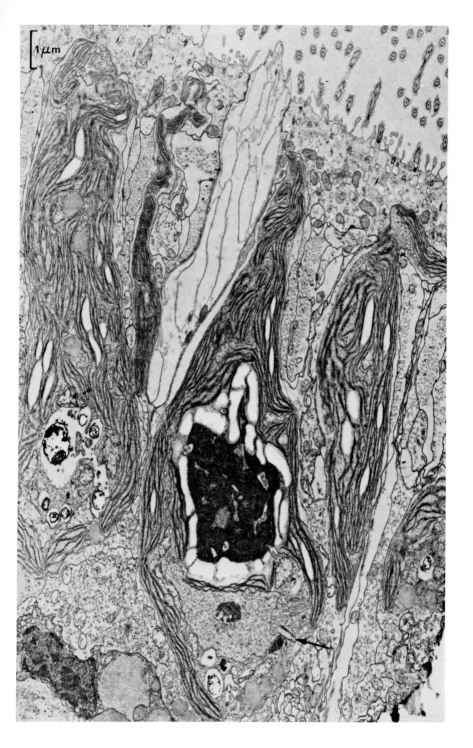

isolated algae only release these in the presence of cellular juices of appropriate animals.[181, 258] The symbiotic algae provide energy, organic carbon, organic nitrogen and oxygen that the animal may use, and obtain carbon dioxide, ammonia, salts and shelter from the host. One indication of the benefit to the animal of the association is given by the observation that corals deposit their calcareous skeletons ten times as quickly in the light as in the dark.

Several freshwater organisms, including the ciliates *Paramecium bursaria* and *Stentor polymorphus*, as well as the coelenterate *Hydra viridis*, contain the green alga *Chlorella*; a single *P. bursaria* may contain hundreds of these zoochlorellae. *P. bursaria* that contain zoochlorellae grow better when provided with a low concentration of bacterial food than do ciliates that lack the algae; it is also interesting that well-fed *P. bursaria* show an increase in the number of zoochlorella cells in the dark, so there is evidently a two-way flow of nutrients. In studies of the relationship between *Hydra* and zoochlorellae, it was found that the alga secretes maltose, especially at acid pH, and that the *Hydra* has developed a maltase to make use of this source of food. Another green alga *Scenedesmus* has been found living in *P. bursaria*. The green flagellate *Platymonas* lives within the cells of the marine flatworm *Convoluta roscoffensis* (Fig. 11.14), while the related flatworm *Convoluta convoluta* contains a symbiotic diatom.

Appendix 1

CLASSIFICATION OF THE PROTOZOA

A newly revised classification of the Protozoa, devised by a Committee of experts led by N. D. Levine, was published in 1980 in the *Journal of Protozoology*, volume 27, pp. 37–58. This regards the Protozoa as composed of seven phyla. Three of these are the groups Sarcomastigophora, Ciliophora and Sporozoa (as Apicomplexa), regarded in this book as sub-phyla. The Cnidospora described here are subdivided by the Committee into three phyla: Myxozoa (comprising members of the Class Myxosporea of this book), Microspora (comprising members of the Order Microsporida of this book) and Ascetospora (comprising members of the Order Haplosporida of this book). The seventh phylum is the Labyrinthulida, a small group of curious motile amoeboid forms, many of which move within a common slime sheath. Within these seven phyla, only the subdivisions of amoebae and ciliates are substantially changed from those given here, but there are already disputes about the new classification of these groups, so in order to avoid confusion the outline below will not be changed for the time being.

An outline of the classification of the phylum Protozoa used in this book, to the level of orders.

Subphylum I. SARCOMASTIGOPHORA

Class I. *MASTIGOPHOREA* (Names of botanical classes shown in parentheses)
Order 1. Cryptomonadida (Cryptophyceae)
2. Haptomonadida (Haptophyceae)
3. Chrysomonadida (Chrysophyceae)
4. Bicoecida
5. Choanoflagellida
6. Silicoflagellida
7. Ebriida
8. Xanthomonadida (Xanthophyceae)
9. (Eustigmatophyceae)
10. Chloromonadida (Rhaphidiophyceae)
11. Dinoflagellida (Dinophyceae)
12. Euglenida (Euglenophyceae)
13. (Prasinophyceae)
14. Phytomonadida [= Volvocida] (Chlorophyceae)
15. Kinetoplastida
16. Metamonadida [= Retortamonadida + Diplomonadida + Oxymonadida + Trichomonadida + Hypermastigida]

Class II. *OPALINATEA*
Order 1. Opalinida

Class III. *SARCODINEA*
Subclass (i) Rhizopodia
Order 1. Amoebida
2. Testacida
Subclass (ii) Granuloreticulosia
Order 1. Foraminiferida
Subclass (iii) Mycetozoia
Order 1. Acrasida
Subclass (iv) Actinopodia
Order 1. Heliozoida
2. Acantharida
3. Radiolarida

Subphylum II. CILIOPHORA
Class I. *CILIATEA*
Subclass (i) Holotrichia
Order 1. Gymnostomatida
2. Trichostomatida
3. Chonotrichida
4. Apostomatida
5. Astomatida
6. Hymenostomatida
7. Thigmotrichida
Subclass (ii) Peritrichia
Order 1. Peritrichida
Sublcass (iii) Suctoria
Order 1. Suctorida
Subclass (iv) Spirotrichia
Order 1. Heterotrichida
2. Oligotrichida
3. Tintinnida
4. Entodiniomorphida
5. Odontostomatida
6. Hypotrichida

Subphylum III. SPOROZOA (= APICOMPLEXA)
Class I. *TELOSPOREA*
Subclass (i) Gregarinia
Order 1. Archigregarinida
2. Eugregarinida
3. Neogregarinida
Subclass (ii) Coccidia
Order 1. Protococcida
2. Eucoccida

Class II. *PIROPLASMEA*
Order 1. Piroplasmida

Subphylum IV. CNIDOSPORA
  Class I. *MYXOSPOREA*
    Order 1. Myxosporida
        2. Actinomyxida
        3. Helicosporida
  Class II. *MICROSPOREA*
    Order 1. Microsporida
        2. Haplosporida

Appendix 2

METHODS OF OBTAINING AND CULTURING PROTOZOA

Protozoa may be collected from the natural sources indicated in many parts of this book, or cultures of organisms may be bought from suppliers and culture collections. Most biological supply houses list a range of the more commonly studied species of free-living Protozoa in their catalogues. A wider range of species is available in most countries from specialized culture collections, (see p. 288), and a list of culture sources and methods has been published in the *Journal of Protozoology* (vol. 5, p. 1, 1958).

A very wide range of culture techniques and recipes for media have been used and will be found in the literature; many of these can be obtained from the culture list in the *Journal of Protozoology*, in appropriate textbooks, [81, 165] and in accounts of studies on the particular species involved. Culture methods for parasitic Protozoa are necessarily rather specific; they will not be mentioned further here, but may be learned from sources mentioned above. Free-living Protozoa may also be cultured by a variety of techniques, but the use of two or three of those listed below will provide large numbers of organisms for study from samples obtained from terrestrial, freshwater or marine habitats. The same culture methods may be used to maintain cultures of selected species isolated from these collections. The maintenance of pure cultures depends upon the food required by the protozoan; generally one has to be content with cultures in which the desired organism is a dominant member of the community of organisms in the culture—most protozoa feed on other living organisms and seldom is a single food organism as satisfactory as a mixed diet. In a few cases axenic or monoxenic culture methods are successful but only a few of the methods mentioned below are suitable.

Samples obtained from freshwater or marine sources contain more Protozoa if they include tufts of filamentous weed, or mud or sand with abundant diatoms or decaying organic matter; a sample of water lacking any substrate is seldom very productive unless the water has a high organic content. It is often worthwhile to make several collections from one body of water, keeping separate containers for samples of mud, weeds of different sorts and the surface film (expecially if there are floating weeds or scum). Similarly with terrestrial samples—dry soil contains spores, but a wider variety of species may usually be obtained from damp soil, particularly if humus is present, and especially if there is a good growth of vegetation; leaf mould and moss are good sources of certain species. Methods for soil Protozoa have recently been described by Heal.[100a]

In order to encourage the multiplication of a variety of species from the samples, it is advisable to separate a sample into several containers and treat each differently. One should be put in a well-illuminated place out of direct sunlight to encourage the growth of photosynthetic species and those that feed upon them; the growth of the plants may be further promoted by addition of a suitable mixture of inorganic salts—obtained most simply by autoclaving 1000 g of garden soil with 1000 cm^3 of water for 30 minutes, adding a little $CaCO_3$ and filtering, or by dissolving the following in 1 litre of distilled water: $NaNO_3$ 250 mg, $CaCO_3$ 25 mg, $MgSO_4$. $7H_2O$ 85 mg, K_2HPO_4 85 mg, KH_2PO_4 185 mg, NaCl 25 mg and a trace of $FeCl_3$. For other cultures less light is advisable but darkness is not normally necessary; in these a variety of bacteria may be encouraged to multiply by providing different food in different dishes—for example the following provide rather different mixtures of Protozoa from the same samples: 0·5% proteose peptone, enough milk to impart a faint turbidity to the water, a few wheat grains that have been boiled to kill the seeds, 20% of hay infusion. The last is made by boiling 10 g of chopped hay in 1 litre of water and allowing the mixture to stand for 24 hours in a flask stoppered with cotton wool before decanting the infusion fluid. The water used for culturing may be boiled tapwater if this is known not to be toxic, but boiled rain or pond water may be substituted, or a solution like Chalkleys medium (below) may be used (boiled seawater for marine species).

The resulting cultures will contain a diversity of amoebae, ciliates and flagellates which may be visible under a dissecting microscope, but usually require a good compound microscope, preferably with phase contrast optics, for adequate study and subsequent identification. Most free living Protozoa are best studied alive, and where there are ciliates that move too quickly to be seen, it is advisable to place a drop of culture within a ring of 2% methyl cellulose and gently lower a cover glass, so that the ciliates will swim slowly in the viscous material. 0·01% nickel sulphate is an effective anaesthetic for many ciliate Protozoa. Many flagellates may be successfully studied after being killed with Lugol's iodine solution (dissolve 4 g of iodine in 100 cm³ of a 6% solution of potassium iodide), and there are special staining methods for parasitic forms, e.g. Giemsa's stain for blood parasites,[9] and silver staining techniques for revealing the infraciliature of ciliates[36, 266]. An extensive list of culture and staining techniques is given in the book by Mackinnon and Hawes.[165]

The correct identification of Protozoa is seldom an easy matter, but some information is to be found in many books. General books on the identification of organisms from particular habitats may have sections on Protozoa, and several chapters in the second edition of Ward and Whipple's *Fresh-Water Biology* may be found especially helpful[55, 146, 189, 259]; the Jahns' guide to the identification of Protozoa[124] is useful, and the systematic sections of many textbooks, especially Kudo[145] and Grassé[90] are also valuable. These books will also provide references to works on individual groups of Protozoa, such as Kahl[130] on ciliates and Cash and Wailes[31] on rhizopods and Heliozoa; some older books like the 3-volume work on Infusoria by Saville Kent[132] are still valuable for their careful illustrations. Information about parasitic Protozoa that may be encountered is also found in general textbooks on Protozoa, as well as in books devoted specifically to the parasitic forms, e.g. the recent book by Baker,[9] as well as monographs on individual groups referred to in various sections of this book. Publications on free-living or parasitic Protozoa of specific habitats can also be most valuable, e.g. ciliates of polluted waters[45] or Protozoa found in British wild mammals.[42]

The culture techniques mentioned above are usually satisfactory for short-term maintenance of mixed cultures of Protozoa, but it is advisable to remove predators and competitors in order to prolong the life of cultures. The following examples of culture techniques used on four common species at Bristol indicate methods that may be used or modified for culture of many other Protozoa, and to provide food organisms for such predatory species as *Didinium*, *Actinosphaerium* and suctorians.

Amoeba proteus is grown in 15 cm diameter glass bowls with glass covers, containing Chalkley's medium (below) to a depth of about 3 cm with 6 boiled wheat grains and 3 boiled 5 cm lengths of Timothy hay stalks. The dish is seeded with amoebae together with flagellates and ciliates for food from an existing culture, and then put in a dimly illuminated place. The amoebae are found mainly on the glass surface, and will persist for many months if more wheat and grass are added monthly. (Chalkley's medium contains the following dissolved in 1 litre of glass distilled water: NaCl 80 mg, $NaHCO_3$ 4 mg, KCl 4 mg, $CaCl_2$ 4 mg, $CaH_4 (PO_4)_2H_2O$, 1·6 mg).

Euglena spp. are grown in jam jars with glass covers three quarters filled with boiled pond water or Chalkley's medium, and with 20 boiled wheat grains; the jars are placed in a well-illuminated situation, but not in direct sunlight. Wheat grains may be added monthly to maintain the cultures.

Paramecium caudatum are grown in jam jars with glass covers, three quarters filled with Chalkley's medium or boiled pond water, and with 7–12 drops of skimmed milk (reconstituted dried milk works well) added weekly. The jars are kept in a dark corner away from direct lighting. Wheat grains or hay extract also provide suitable food for the bacteria that *Paramecium* eat.

Tetrahymena pyriformis are kept in axenic cultures in rimless 30 ml test tubes with metal caps that are autoclaved after addition of 15 cm^3 of peptone medium (10 g proteose peptone, 5 g NaCl in 1 litre of distilled water); subcultures are made every 3–4 weeks using sterile pipettes and normal sterile precautions, and the cultures are kept in the dark. Axenic cultures of the flagellates *Chilomonas* and *Astasia* are maintained in more dilute peptone media in which the NaCl is replaced with sodium acetate.

Cultures may be obtained from the following :

The Culture Collection of Algae and Protozoa, 36 Storey's Way, Cambridge, CB3 0DT, England.

Algensammlung, Pflanzenphysiologisches Institut, Göttingen, West Germany.

The Curator of the Culture Collection of Algae, Botany Department, Indiana University, Bloomington, Indiana, 47401, U.S.A.

American Type Culture Collection, 12301 Parklawn Drive, Rockville, Maryland, 20852, U.S.A.

References

1. AIELLO, E. and SLEIGH, M. A. (1972). The metachronal wave of lateral cilia of *Mytilus edulis*. *J. Cell Biol.*, **54**, 493–506.
2. ALLEN, RICHARD D. (1967). Fine structure, reconstruction and possible functions of components of the cortex of *Tetrahymena pyriformis*. *J. Protozool.*, **14**, 553–565.
3. ALLEN, RICHARD D. (1969). The morphogenesis of basal bodies and accessory structures of the cortex of the ciliated protozoon *Tetrahymena pyriformis*. *J. Cell Biol.*, **40**, 716–733.
4. ALLEN, RICHARD D. (1971). Fine structure of membranous and microfibrillar systems in the cortex of *Paramecium caudatum*. *J. Cell Biol.*, **49**, 1–20.
5. ALLEN, ROBERT D. (1968). Differences of a fundamental nature among several types of amoeboid movement. *Symp. Soc. exp. Biol.*, **22**, 151–168.
6. ALLEN, ROBERT D. (1970). Comparative aspects of amoeboid movement. *Acta Protozool.*, **7**, 291–299.
7. AMOS, W. B. (1971). Reversible mechanochemical cycle in the contraction of *Vorticella*. *Nature, Lond.*, **229**, 127–128.
8. AMOS, W. B. (1972). Structure and coiling of the stalk in the peritrich ciliates *Vorticella* and *Carchesium*. *J. Cell Sci.*, **10**, 95–122.
9. BAKER, J. R. (1969). *Parasitic Protozoa*. Hutchinson, London.
10. BALL, G. H. (1969). Organisms living on and in Protozoa. pp. 565–718. In *Research in Protozoology*, **3**, CHEN, T.-T. Pergamon Press, Oxford.
11. BANNISTER, L. H. and TATCHELL, E. C. (1968). Contractility and the fibre system of *Stentor coeruleus*. *J. Cell Sci.*, **3**, 295–308.
12. BARDELE, C. F. (1969). Ultrastruktur der Körnchen auf den Axopodien von *Raphidiophrys* (Centrohelida, Heliozoa). *Z. Naturforsch.*, **24b**, 362–363.
13. BARDELE, C. F. (1970). Budding and metamorphosis in *Acineta tuberosa*. An electron microscope study on morphogenesis in Suctoria. *J. Protozool.*, **17**, 51–70.
14. BARDELE, C. F. (1971). Microtubule model systems: cytoplasmic transport in the suctorian tentacle and the centrohelidian axopod. *Proc. 29 A. Meet. Elect. Micr. Soc. Am.*, 334–335.

15. BARDELE, C. F. (1972). A microtubule model for ingestion and transport in the suctorian tentacle. *Z. Zellforsch.*, **126**, 116–134.
16. BARDELE, C. F. (1972). Cell cycle, morphogenesis, and ultrastructure in the pseudoheliozoan *Clathrulina elegans*. *Z. Zellforsch.*, **130**, 219–242.
17. BARDELE, C. F. and GRELL, K. G. (1967). Elektronenmikroskopische Beobachtungen zur Nahrungsaufnahme bie dem Suktor *Acineta tuberosa* Ehrenberg. *Z. Zellforsch.*, **80**, 108–123.
18. BAUDHUIN, P. (1969). Peroxisomes (microbodies, glyoxisomes). pp. 1179–1195. In *Handbook of Molecular Cytology*, LIMA-DE-FARIA, A., North Holland, Amsterdam.
19. BEALE, G. H. (1954). *The Genetics of Paramecium aurelia.* Cambridge University Press.
20. BEAMS, H. W. and SEKHON, S. S. (1969). Ultrastructure of *Lophomonas blattarum. J. Ultrastruct. Res.*, **26**, 296–315.
21. BĚLAŘ, K. (1926). Der Formwechsel der Protistenkerne. *Ergebn. Zool.*, **6**, 235–654.
21a. BICK, H. (1967). Vergleichende Untersuchung der Ciliatensukzession beim Abbau von Pepton und Cellulose (Modelleversuche). *Hydrobiologia*, 30, 353–373.
22. BONNER, J. T. (1967). *The Cellular Slime Molds*, 2nd Edn. Princeton University Press, Princeton, N.J.
23. BORROR, A. C. (1963). Morphology and ecology of the benthic ciliated Protozoa of Alligator Harbor, Florida. *Arch. Protistenk.*, **106**, 465–534.
24. BROBYN, P. (1970). Studies on feeding of *Didinium* on *Paramecium*. 'Unpublished'.
25. BROKAW, C. J. (1965). Non-sinusoidal bending waves of sperm flagella. *J. exp. Biol.*, **43**, 155–169.
26. BUETOW, D. E. (1968). *The Biology of* Euglena. 2 vols. Academic Press, New York.
27. CACHON, J. and CACHON-ENJUMET, M. (1967). Contribution à l'étude des Noctilucidae (Saville-Kent). I. Les Kofoidininae (Cachon, J. et M.), évolution morphologique et systématique. *Protistologica*, **3**, 427–444.
28. CACHON, J. and CACHON, M. (1969). Contribution à l'étude des Noctilucidae (Saville-Kent). Évolution morphologique, cytologie, systematique. II. Les Leptodiscinae. *Protistologica*, **5**, 11–33.
29. CACHON, J. and CACHON, M. (1971). The axopodial system of Radiolaria Nassellaria. Origin, Organization and relation with the other Cell Organelles. General considerations on the macromolecular organization of the stereoplasm in Actinopods. *Arch. Protistenk.*, **113**, 80–97.
29a. CACHON, J. and CACHON, M., (1972) Les modalites du depot de la Silice chez les Radiolaires. *Arch. Protistenk.*, **14**, 1–13.
30. CALKINS, G. N. and SUMMERS, F. M. (1941). *Protozoa in Biological Research.* Columbia University Press, New York.
31. CASH, J. and WAILES, G. H. (1905–1921). *The British Freshwater Rhizopoda and Heliozoa.* 5 vols. Ray Society, London.
32. CHEN, T-T. (1967–1972). *Research in Protozoology.* 4 vols. Pergamon Press, Oxford.
33. CLEVELAND, L. R. (1950). Hormone-induced sexual cycles of flagellates. II Gametogenesis, Fertilization, and one-division meiosis in *Oxymonas. J. Morph.*, **86**, 185–214.

34. CLEVELAND, L. R. (1956). Brief accounts of the sexual cycles of the flagellates of *Cryptocercus*. *J. Protozool.*, **3**, 161–180.
35. CLEVELAND, L. R. and GRIMSTONE, A. V. (1964). The fine structure of the flagellate *Mixotricha paradoxa* and its associated micro-organisms. *Proc. R. Soc. Lond. B*, **159**, 668–685.
36. CORLISS, J. O. (1953). Silver impregnation of ciliated Protozoa by the Chatton-Lwoff technic. *Stain Technol.*, **28**, 97–100.
37. CORLISS, J. O. (1955). The Opalinid infusorians: flagellates or ciliates? *J. Protozool.*, **2**, 107–114.
38. CORLISS, J. O. (1959). An illustrated key to the higher groups of ciliated Protozoa, with definition of terms. *J. Protozool.*, **6**, 265–281.
39. CORLISS, J. O. (1959). Comments on the systematics and phylogeny of the Protozoa. *Syst. Zool.*, **8**, 169–190.
40. CORLISS, J. O. (1961). *The Ciliated Protozoa: Characterization, Classification and Guide to the Literature*. Pergamon Press, Oxford.
41. CORLISS, J. O. (1965). *Tetrahymena*, a ciliate genus of unusual importance in modern biological research. *Acta Protozool.*, **3**, 1–20.
41a. CORLISS, J. O. (1972). The ciliate Protozoa and other organisms: some unresolved questions of major phylogenetic significance. *Am. Zoologist*, 12, 739–753.
42. COX, F. E. G. (1970). Parasitic Protozoa of British wild mammals. *Mamm. Rev.*, **1**, 1–28.
43. COX, F. E. G. and VICKERMAN, K. (1966). Pinocytosis in *Plasmodium vinckei*. *Ann. trop. Med. Parasit.*, **60**, 293–296.
44. CURDS, C. R. (1966). An ecological study of the ciliated Protozoa in activated sludge. *Oikos*, **15**, 282–289.
45. CURDS, C. R. (1969). *An Illustrated Key to the British Freshwater Ciliated Protozoa commonly found in Activated Sludge*. H.M.S.O., London.
45a. CURDS, C. R. (1971). Computer simulations of microbial population dynamics in the activated-sludge process. *Wat. Res.*, 5, 1049–1066.
46. CURDS, C. R. and COCKBURN, A. (1968). Studies on the growth and feeding of *Tetrahymena pyriformis* in axenic and monoxenic culture. *J. gen. Microbiol.*, **54**, 343–358.
47. CURDS, C. R. and COCKBURN, A. (1970). Protozoa in biological sewage-treatment processes—I. A survey of the protozoan fauna of British percolating filters and activated-sludge plants. *Wat. Res.*, **4**, 225–236.
48. CURDS, C. R. and COCKBURN, A. (1970). Protozoa in biological sewage-treatment processes—II. Protozoa as indicators in the activated-sludge process. *Wat. Res.*, **4**, 237–249.
49. CURDS, C. R. and COCKBURN, A. (1971). Continuous monoxenic culture of *Tetrahymena pyriformis*. *J. gen. Microbiol.*, **66**, 95–108.
50. CURDS, C. R. COCKBURN, A. VANDYKE, J. M. (1968). An experimental study of the role of the ciliated Protozoa in the activated-sludge process. *Wat. Pollut. Control.* **67**, 312–329.
51. CURDS, C. R. and VANDYKE, J. M. (1966). The feeding habits and growth rates of some fresh-water ciliates found in activated-sludge plants. *J. appl. Ecol.*, **3**, 127–137.
52. DANFORTH, W. F. (1967). Respiratory Metabolism. pp. 201–306. In *Research in Protozoology*, **1**, CHEN, T.-T., Pergamon Press, Oxford.
53. DARBYSHIRE, J. F. (1966). Protozoa in the rhizosphere of *Lolium perenne* L. *Can. J. Microbiol.*, **12**, 1287–1289.
54. DAVIES, S. F. M., JOYNER, L. P. and KENDALL, B. S. (1963). *Coccidiosis*. Oliver and Boyd, Edinburgh.

55. DEFLANDRE, G. (1959). Rhizopoda and Actinopoda. pp. 232–264. In *Fresh-Water Biology*, EDMONDSON, W. T., Wiley, New York.
56. DE TERRA, N. (1970). Cytoplasmic control of macronuclear events in the cell cycle of *Stentor*. *Symp. Soc. exp. Biol.*, **24**, 345–368.
57. DINGLE, A. D. and FULTON, C. (1966). Development of the flagellar apparatus of *Naegleria*. *J. Cell Biol.*, **31**, 43–54.
58. DODGE, J. D. (1963). The nucleus and nuclear division in the Dinophyceae. *Arch. Protistenk.*, **106**, 442–452.
59. DODGE, J. D. (1968). The fine structure of chloroplasts and pyrenoids in some marine dinoflagellates. *J. Cell Sci.*, **3**, 41–48.
60. DODGE, J. D. (1969). The ultrastructure of *Chroomonas mesostigmatica* Butcher (Cryptophyceae). *Arch. Mikrobiol.*, **69**, 266–280.
60a. DODGE, J. D. (1971). Fine structure of the Pyrrophyta. *Bot. Rev.*, **37**, 484–508.
61. DODGE, J. (1972). The ultrastructure of the dinoflagellate pusule: a unique osmo-regulatory organelle. *Protoplasma*, **75**, 285–302.
62. DODGE, J. D. and CRAWFORD, R. M. (1968). Fine structure of the dinoflagellate *Amphidinium carteri*. Hulbert. *Protistologica*, **4**, 231–242.
63. DODGE, J. D. and CRAWFORD, R. M. (1969). Observations on the fine structure of the eyespot and associated organelles in the dinoflagellate *Glenodinium foliaceum*. *J. Cell Sci.*, **5**, 479–493.
64. DODGE, J. D. and CRAWFORD, R. M. (1970). A survey of thecal fine structure in the Dinophyceae. *Bot. J. Linn. Soc.*, **63**, 53–67.
65. DODGE, J. D. and CRAWFORD, R. M. (1970). The morphology and fine structure of *Ceratium hirudinella* (Dinophyceae). *J. Phycol.*, **6**, 137–149.
66. DODGE, J. D. and CRAWFORD, R. M. (1971). Fine structure of the dinoflagellate *Oxyrrhis marina*. *Protistologica*, **7**, 399–409.
67. DOGIEL, V. A. (1965). *General Protozoology*. Revised by J. I. POLJANSKIJ and E. M. CHEJSIN. 2nd Edn. University Press, Oxford.
68. DRAGESCO, J. (1963). Compléments à la connaissance des Ciliés mésopsammiques de Roscoff. 1. Holotriches. *Cah. Biol. mar.*, **4**, 91–119.
69. DRAGESCO, J. (1963). Compléments à la connaissance des Ciliés mésopsammiques de Roscoff. II. Hétérotriches. III. Hypotriches. *Cah. Biol. mar.*, **4**, 251–275.
70. DRYL, S. (1963). Oblique galvanotaxis in *Stylonychia mytilus*. *J. Protozool.*, **10**, suppl. 26.
71. DUBEY, J. P. and FRENKEL, J. K. (1972). Cyst-induced toxoplasmosis in cats. *J. Protozool*, **19**, 155–177.
72. ECKERT, R. (1972). Bioelectric control of ciliary activity. *Science, N.Y.*, **176**, 473–481.
73. ETTIENNE, E. M. (1970). Control of contractility in *Spirostomum* by dissociated calcium ions. *J. gen. Physiol.*, **56**, 168–179.
74. FAURÉ-FREMIET, E. (1951). The marine sand-dwelling ciliates of Cape Cod. *Biol. Bull. mar. biol. Lab.*, *Woods Hole*, **100**, 59–70.
75. FAURÉ-FREMIET, E. (1951). Associations infusoriennes à *Beggiatoa*. *Hydrobiol.*, **3**, 65–71.
75a. FJERDINGSTAD, E. (1971). Microbial criteria of environment qualities. *Ann. Rev. Microbiol.*, **25**, 563–582.
76. FORER, A., NILSSON, J. R. and ZEUTHEN, E. (1970). Studies on the oral apparatus of *Tetrahymena pyriformis* G1. *Compt. Rend. Trav. Lab. Carlsberg*, **38**, 67–86.
77. FOTT, B. (1971). *Algenkunde*. Gustav Fischer, Jena.

78. FRANKEL, J. (1967). Studies on the maintenance of oral development in *Tetrahymena pyriformis* GL-C. II. The relationship of protein synthesis to cell division and oral organelle development. *J. Cell Biol.*, **34**, 841–858.

79. FRIEDHOFF, K. T. and SCHOLTYSECK, E. (1968). Fine Structure of *Babesia ovis* trophozoites in *Rhipicephalus bursa* ticks. *J. Parasit.*, **54**, 1246–1250.

80. FRITSCH, F. E. (1965). *The Structure and Reproduction of the Algae*. 2 vols. Cambridge University Press.

80a. FULTON, C. and DINGLE, A. D. (1971). Basal bodies, but not centrioles, in *Naegleria*. *J. Cell Biol.*, **51**, 826–836.

81. GALTSOFF, P. S., LUTZ, F. E., WELCH, P. S. and NEEDHAM, J. G. (1959). *Culture Methods for Invertebrate Animals*. A.A.A.S., Dover, New York.

82. GARNHAM, P. C. C. (1966). *Malaria Parasites and other Haemosporidia*. Blackwell, Oxford.

83. GAUMONT, R. and GRAIN, J. (1967). L'anaerobiose et les mitochondries chez les protozoaires du tube digestif. *Annls. Univ. A.R.E.R.S.*, *Reims*, **5**, 174–176.

84. GAUSE, G. F. (1934). *The Struggle for Existence*. Williams and Wilkins, Baltimore, Md.

85. GIBBONS, I. R. (1965). Chemical dissection of cilia. *Arch. Biol.*, *Liège*, **76**, 317–352.

86. GIBBS, S. P. (1962). Nuclear envelope-chloroplast relationships of algae. *J. Cell Biol.*, **14**, 433–444.

87. GLIDDON, R. (1966). Ciliary organelles and associated fibre systems in *Euplotes eurystomus* (Ciliata, Hypotrichida) I. Fine structure. *J. Cell Sci.*, **1**, 439–448.

88. GOJDICS, M. (1953). *The genus Euglena*. University of Wisconsin Press, Madison.

89. GRAIN, J. (1969). Le cinétosome et ses dérivés chez les ciliés. *Ann. Biol.*, **8**, 54–97.

90. GRASSÉ, P.-P. (1952). *Traité de Zoologie* **1** (i) and (ii). Masson, Paris.

91. GRĘBECKI, A. (1965). Role of Ca^{2+} ions in the excitability of protozoan cell. Decalcification, recalcification, and the ciliary reversal in *Paramecium caudatum*. *Acta Protozool.*, **3**, 275–289.

92. GRELL, K. G. (1956). *Protozoologie*. Springer, Berlin.

93. GRELL, K. G. (1967). Sexual reproduction in Protozoa. pp. 147–213. In *Research in Protozoology*, **2**, CHEN, T.-T., Pergamon Press, Oxford.

94. GRIMSTONE, A. V. (1961). Fine structure and morphogenesis in Protozoa. *Biol. Rev.*, **36**, 97–150.

95. GRIMSTONE, A. V. and CLEVELAND, L. R. (1965). The fine structure and function of the contractile axostyles of certain flagellates. *J. Cell Biol.*, **24**, 387–400.

96. HADZI, J. (1963). *The Evolution of the Metazoa*. Pergamon Press, Oxford.

97. HALL, R. P. (1967). Nutrition and growth of Protozoa, pp. 337–404. In *Research in Protozoology*, **1**. CHEN, T.-T. Pergamon Press, Oxford.

98. HALLDAL, P. (1964). Phototaxis in Protozoa, pp. 277–296. In *Biochemistry and Physiology of Protozoa*, **3**, HUTNER, S. H., Academic Press, New York.

99. HAUSMANN, K., STOCKEM, W. and WOHLFARTH-BOTTERMANN, K. E. (1972). Cytological studies on trichocysts. II. The fine structure of resting

and inhibited spindle trichocysts in *Paramecium caudatum*. *Cytobiol.*, **5**, 228–246.

100. HAWES, R. S. J. (1963). The emergence of asexuality in Protozoa. *Quart Rev. Biol.*, **38**, 234–242.

100a. HEAL, O. W. (1971). Protozoa, pp. 51–71 in *Quantitative soil ecology*. Phillipson, J., (I.B.P. Handbook 18) Blackwell, Oxford.

101. HEDLEY, R. H. (1964). The biology of Foraminifera. *Int. Rev. gen. exp. Zool.*, **1**, 1–45.

101a. HEDLEY, R. H. and OGDEN, C. G. (1973). Biology and fine structure of *Euglypha rotunda* (Testacea: Protozoa). *Bull. Br. Mus. nat. Hist.* (*Zool.*) **25**, 121–137.

102. HEDLEY, R. H. and WAKEFIELD, J. ST. J. (1969). Fine structure of *Gromia oviformis* (Rhizopodea: Protozoa). *Bull. Br. Mus. nat. Hist.* (*Zool.*), **18**, 69–89.

102a. HEYWOOD, P. (1972). Structure and origins of flagellar hairs in *Vacuolaria virescens*. *J. Ultrastruct. Res.*, **39**, 608–623.

103. HIBBERD, D. J. and LEEDALE, G. F. (1972). Observations on the cytology and ultrastructure of the new algal class, Eustigmatophyceae. *Ann. Bot.*, **36**, 49–71.

104. HOARE, C. A. (1972). *The Trypanosomes of Mammals: a Zoological Monograph*. Blackwell, Oxford.

105. HOLLANDE, A., CACHON, J. and CACHON, M. (1969). La dinomitose atractophorienne à fuseau endonucléaire chez les Radiolaires Thalanophysidae. Son homologie avec la mitose des Foraminifères et avec celles des levures. *C.r. hebd. Séanc. Acad. Sci.*, *Paris*, **269**, 179–182.

106. HOLLANDE, A., CACHON, J., CACHON-ENJUMET, M. and VALENTIN, J. (1967). Infrastructure des axopodes et organisation générale de *Sticholonche zanclea* (Hertwig) (Radiolaire Sticholonchidae). *Protistologica*, **3**, 155–166.

107. HOLLANDE, A. and VALENTIN, J. (1967). Morphologie et infrastructure du genre *Barbulanympha*, Hypermastigine Symbiontique de *Cryptocercus punctulatus* Scudder. *Protistologica*, **3**, 257–267.

108. HOLLANDE, A. and VALENTIN, J. (1968). Morphologie infrastructurale de *Trichomonas* (*Trichomitopsis* Kofoid et Swezy 1919) *termopsidis*, parasite intestinal de *Termopsis angusticollis* Walk. Critique de la notion de centrosome chez les polymastigines. *Protistologica*, **4**, 127–140.

109. HOLLANDE, A. and VALENTIN, J. (1968). Infrastructure du complexe rostral et origine du fuseau chez *Staurojoenina caulleryi*. *C.r. hebd. Séanc. Acad. Sci.*, *Paris*, **266**, 1283–1286.

110. HOLLANDE, A. and VALENTIN, J. (1969). Appareil de Golgi, pinocytose, lysosomes, mitochondries, bactéries symbiontiques, atractophores et pleuromitose chez les hypermastigines du genre *Joenia*. Affinités entre Joenides et Trichomonadines. *Protistologica*, **5**, 39–86.

111. HOLWILL, M. E. J. (1964). The motion of *Strigomonas oncopelti*. *J. exp. Biol.*, **42**, 125–137.

112. HOLWILL, M. E. J. (1966). Physical aspects of flagellar movement. *Physiol. Rev.*, **46**, 696–785.

113. HOLWILL, M. E. J. and SLEIGH, M. A. (1967). Propulsion by hispid flagella. *J. exp. Biol.*, **47**, 267–276.

114. HOLZ, G. G. (1960). Structural and functional changes in a generation in *Tetrahymena*. *Biol. Bull. mar. biol. Lab.*, *Woods Hole*, **118**, 84–95.

115. HONIGBERG, B. M., BALAMUTH, W., BOVEE, E. C., CORLISS, J. O., GOJDICS, M., HALL, R. P., KUDO, R. R., LEVINE, N. D., LOEBLICH, A. R., JNR., WEISER, J., and WENRICH, D. H. (1964). A revised classification of the phylum Protozoa. *J. Protozool.*, **11**, 7–20.

116. HUANG, B. (1970). Ultrastructure of the cortical fiber systems in *Stentor coeruleus* relaxed in E.G.T.A. *J. Cell Biol.*, **47**, 92a.

117. HUNGATE, R. E. (1955). Mutualistic intestinal Protozoa, p. 159. In *Biochemistry and Physiology of Protozoa*, **2**, HUTNER, S. H. and LWOFF, A., Academic Press, New York.

118. HUTCHISON, W. M., DUNACHIE, J. F., SIIM, J. C. and WORK, K. (1970). Coccidian-like nature of *Toxoplasma gondii*. *Brit. Med. J.*, **1**, 142–144.

118a. HUTNER, S. H. (1964). Editor. *Biochemistry and Physiology of Protozoa*, **3**. Academic Press, New York.

118b. HUTNER, S. H. and LWOFF, A. (1955). Editors. *Biochemistry and Physiology of Protozoa*, **2**. Academic Press, New York.

119. HYMAN, L. H. (1940). *The Invertebrates I: Protozoa through Ctenophora*, McGraw-Hill, New York.

120. HYNES, H. B. N. (1960). *The Biology of Polluted Waters*. Liverpool University Press.

121. ISSI, I. V. and SHULMAN, S. S. (1968). The systematic position of Microsporidia. *Acta Protozool.*, **6**, 121–135.

122. JAHN, T. L. (1962). The mechanism of ciliary movement II. Ion antagonism and reversal. *J. cell. comp. Physiol.*, **60**, 217–228.

123. JAHN, T. L. and BOVEE, E. C. (1967). Motile behaviour of Protozoa, pp. 41–200. In *Research in Protozoology*, **1**, CHEN, T.-T., Pergamon Press, Oxford.

124. JAHN, T. L. and JAHN, F. F. (1949). *How to Know the Protozoa*, Brown, Dubuque, Iowa.

125. JAHN, T. L. and RINALDI, R. A. (1959). Protoplasmic movement in the foraminiferan, *Allogromia laticollaris;* and a theory of its mechanism. *Biol. Bull. mar. biol. Lab.*, *Woods Hole*, **117**, 100–118.

126. JEPPS, M. W. (1956). *The Protozoa, Sarcodina*. Oliver and Boyd, Edinburgh.

127. JONES, A. R., JAHN, T. L. and FONSECA, J. R. (1970). Contraction of protoplasm IV. Cinematographic analysis of the contraction of some peritrichs. *J. cell Physiol.*, **75**, 9–20.

128. JØRGEN HANSEN, H. (1972). Pore pseudopodia and sieve plates of *Amphistegina*. *Micropalaeont.*, **18**, 223–230.

129. JURAND, A. and SELMAN, G. G. (1969). *The Anatomy of Paramecium aurelia*, Macmillan, London.

130. KAHL, A. (1932). Protozoa: Ciliata, part 25 In *Die Tierwelt Deutschlands*, DAHL, F., Gustav Fischer, Jena.

131. KENNEDY, J. R. (1965). The morphology of *Blepharisma undulans* Stein. *J. Protozool.*, **12**, 542–561.

132. KENT, W. SAVILLE. (1880–1). *A manual of the Infusoria*. 3 vols., Bogue, London.

133. KERKUT, G. A. (1960). *Implications of Evolution*. Pergamon Press, Oxford.

134. KHAN, M. A. (1969). Fine structure of *Ancistrocoma pelseneeri* (Chatton et Lwoff), a rhynchodine thigmotrichid ciliate. *Acta Protozool.*, **7**, 29–47.

135. KIMBALL, R. (1942). The nature and inheritance of mating types in *Euplotes patella. Genetics*, **27**, 269–285.
136. KINOSITA, H. (1954). Electric potentials and ciliary response in *Opalina. J. Fac. Sci. Tokyo Univ. Sec IV*, **7**, 1–14.
137. KINOSITA, H. and MURAKAMI, A. (1967). Control of ciliary motion. *Physiol. Rev.*, **47**, 53–82.
138. KIRBY, H. (1941). Relationships between certain Protozoa and other animals, pp. 890–1008. In *Protozoa in Biological Research*, CALKINS, G. N. and SUMMERS, F. M. Columbia University Press, New York.
139. KITCHING, J. A. (1951). The physiology of contractile vacuoles VII. Osmotic relations in a suctorian, with special reference to the mechanism of control of vacuolar output. *J. exp. Biol.*, **28**, 203–214.
140. KITCHING, J. A. (1956). Contractile vacuoles of Protozoa. *Protoplasmatologia*, **III**, D 3a, 1–45.
141. KITCHING, J. A. (1956). Food vacuoles of Protozoa. *Protoplasmatologia*, **III**, D 3b, 1–54.
142. KITCHING, J. A. (1957). Some factors in the life of free-living Protozoa. *Symp. Soc. gen. Microbiol.*, **7**, 259–286.
142a. KITCHING, J. A. (1964). The axopods of the Sun Animalcule, *Actinophrys sol* (Heliozoa), pp. 445–456. In *Primitive Motile Systems in Cell Biology*. ALLEN, R. D. and KAMIYA, A. Academic Press, New York.
143. KITCHING, J. A. (1967). Contractile vacuoles, ionic regulation and excretion. pp. 307–336. In *Research in Protozoology*, **1**, CHEN, T.-T., Pergamon Press, Oxford.
144. KNIGHT-JONES, E. W. (1954). Relations between metachronism and the direction of ciliary beat in Metazoa. *Q. Jl microsc. Sci.*, **95**, 503–521.
145. KUDO, R. R. (1966). *Protozoology*, 5th Edn. Thomas, Springfield, Illinois.
146. LACKEY, J. B. (1959). Zooflagellates, pp. 190–231. In *Fresh-Water Biology*. EDMONDSON, W. T. Wiley, New York.
147. LEADBEATER, B. S. C. (1971). Observations by means of cine photography on the behaviour of the haptonema in plankton flagellates of the Class Haptophyceae. *J. mar. biol. Ass. U.K.*, **51**, 207–217.
148. LEADBEATER, B. S. C. (1972). Fine-structural observations on some marine choanoflagellates from the coast of Norway. *J. mar. biol. Ass. U.K.*, **52**, 67–79.
148a. LEADBEATER, B. S. C. (1972). *Paraphysomonas cylicophora* sp. nov., a marine species from the coast of Norway. *Norw. J. Bot.*, 19, 179–185.
149. LEADBEATER, B. S. C. and DODGE, J. D. (1967). An electron microscope study of nuclear and cell division in a dinoflagellate. *Arch. Mikrobiol.*, **57**, 239–254.
150. LEADBEATER, B. S. C. and DODGE, J. D. (1967). An electron microscope study of dinoflagellate flagella. *J. gen. Microbiol.*, **46**, 305–314.
151. LE CALVEZ, J. (1953). Ordre des Foraminifères, pp. 149–265. In *Traité de Zoologie*, 1 (ii), GRASSE, P.-P., Masson, Paris.
152. LEEDALE, G. F. (1964). Pellicle structure in *Euglena. Br. phycol. Bull.*, **5**, 291–306.
153. LEEDALE, G. F. (1967). *Euglenoid Flagellates*. Prentice-Hall, Englewood Cliffs, New Jersey.
154. LEEDALE, G. F. (1968). The nucleus in *Euglena*, pp. 185–242. In *The Biology of Euglena*, 1, BUETOW, D. E., Academic Press, New York.

155. LEGRAND, B. (1970). Recherches expérimentales sur le détérminisme de la contraction et les structures contractiles chez le spirostome. *Protistologica*, **6**, 283–300.

156. LENGSFELD, A. M. (1969). Nahrungsaufnahme und Verdauung bei der Foraminifere *Allogromia laticollaris*. *Helgoländer wiss. Meeresunters*, **19**, 385–400.

157. LEVINE, N. D. (1969). Taxonomy of the sporozoa. *Proc. 3rd Int. Cong. Protozool.*, Leningrad, 365–367.

158. LEVINE, N. D. (1972). Relationship between certain Protozoa and other animals, pp. 291–350. In *Research in Protozoology*, **4**, CHEN, T.-T., Pergamon Press, Oxford.

159. LOM, J. (1964). The morphology and morphogenesis of the buccal ciliary organelles in some peritrichous ciliates. *Arch. Protistenk.*, **107**, 131–162.

160. LOM, J. and CORLISS, J. O. (1971). Morphogenesis and cortical ultrastructure of *Brooklynella hostilis*, a Dysteriid Ciliate ectoparasitic on marine fishes. *J. Protozool.*, **18**, 261–281.

161. LOM, J., CORLISS, J. O. and NOIROT-TIMOTHÉE, C. (1968). Observations on the ultrastructure of the buccal apparatus in thigmotrich ciliates and their bearing on Thigmotrich-Peritrich affinities. *J. Protozool.*, **15**, 824–840.

162. LWOFF, A. (1950). *Problems of Morphogenesis in Ciliates*, Wiley, New York.

162a. LWOFF, A. (1951). Editor, *Biochemistry and Physiology of Protozoa*, **1**. Academic Press, New York.

163. MACDONALD, A. C. and KITCHING, J. A. (1967). Axopodial filaments of Heliozoa. *Nature, Lond.*, **215**, 99–100.

164. MACHEMER, H. (1972). Properties of polarized ciliary beat in *Paramecium*. *Acta Protozool.*, **11**, 295–300.

165. MACKINNON, D. L. and HAWES, R. S. J. (1961). *An Introduction to the Study of Protozoa*. University Press, Oxford.

166. MADISON, K. M. (1958). Fossil protozoans from the Keewatin sediments. *Trans. Ill. St. Acad. Sci.*, **50**, 287–290.

167. MANTON, I. (1964). Observations with the electron microscope on the division cycle in the flagellate *Prymnesium parvum* Carter. *Jl R. micr. Soc.*, **83**, 317–325.

168. MANTON, I. (1964). Further observations on the fine structure of the haptonema in *Prymnesium parvum*. *Arch. Mikrobiol.*, **49**, 315–330.

169. MANTON, I. (1965). Some phyletic implications of flagellar structure in plants. *Adv. bot. Res.*, **2**, 1–34.

170. MANTON, I. (1968). Further observations on the microanatomy of the haptonema in *Chrysochromulina chiton* and *Prymnesium parvum*. *Protoplasma*, **66**, 35–53.

171. MANTON, I. and LEEDALE, G. F. (1961). Observations on the fine structure of *Paraphysomonas vestita*, with special reference to the Golgi apparatus and the origin of scales. *Phycologia*, **1**, 37–57.

172. MANTON, I. and LEEDALE, G. F. (1969). Observations on the microanatomy of *Coccolithus pelagicus* and *Cricosphaera carterae*, with special reference to the origin and nature of coccoliths and scales. *J. mar. biol. Ass. U.K.*, **49**, 1–16.

173. MARGULIS, L. (1970). *Origin of eukaryotic cells*. Yale University Press, New Haven.

174. MARTIN, D. (1968). *Microfauna of biological filters*. Oriel Press, Newcastle-upon-Tyne.

174a. MARTIN, G. W. and ALEXOPOULOS, C. J. (1969). *The Myxomycetes*. University of Iowa Press, Iowa City.

175. MASSALSKI, A. and LEEDALE, G. F. (1969). Cytology and ultrastructure of the Xanthophyceae. I. Comparative morphology of the zoospores of *Bumilleria sicula* Borzi and *Tribonema vulgare* Pascher. *Br. phycol. J.*, **4**, 159–180.

176. MCGEE-RUSSELL, S. M. and ALLEN, R. D. (1971). Reversible stabilization of labile microtubules in the reticulopodial network of *Allogromia*. *Adv. Cell molec. Biol.*, **1**, 153–184.

176a. MCINTOSH, J. R., CLELAND, S. and OGATA, E. S. (1970). Motion and structure of the axostyle in *Saccinobaculus*. *J. Cell Biol.*, **47**, 134a.

177. MERCER, E. H. (1959). An electron microscope study of *Amoeba proteus*. *Proc. R. Soc. Lond.* B, **150**, 216–232.

178. MÜLLER, M. and HOGG, J. F. (1967). Occurrence of protozoal isocitrate lyase and malate synthetase in the peroxisomes. *Fedn Proc. Fedn Am. Socs exp. Biol.*, **26**, 284.

179. MÜLLER, M. and MØLLER, K. M. (1969). Urate oxidase and its association with peroxisomes in *Acanthamoeba* sp. *Eur. J. Bioch.*, **9**, 424–430.

180. MURRAY, J. W. (1971). *An Atlas of British Recent Foraminiferids*. Heinemann, London.

181. MUSCATINE, L. (1967). Glycerol excretion by symbiotic algae from corals and *Tridacna* and its control by the host. *Science, N.Y.*, **156**, 516–519.

182. NAITOH, Y. and ECKERT, R. (1969). Ionic mechanisms controlling behavioural responses of *Paramecium* to mechanical stimulation. *Science, N.Y.*, **164**, 963–965.

183. NAITOH, Y. and ECKERT, R. (1969). Ciliary orientation: Controlled by cell membrane or by intracellular fibrils? *Science, N.Y.*, **166**, 1633–1635.

184. NEFF, R. J. and NEFF, R. H. (1969). The biochemistry of amoebic encystment. *Symp. Soc. exp. Biol.*, **23**, 51–81.

185. NETZEL, H. (1971). Die Schalenbildung bei der Thekamoben—Gattung *Arcella* (Rhizopoda, Testacea). *Cytobiol.*, **3**, 89–92.

186. NEWELL, P. C., FRANKE, J. and SUSSMAN, M. (1972). Regulation of four functionally related enzymes during shifts in the developmental program of *Dictyostelium discoideum*. *J. molec. Biol.*, **63**, 373–382.

187. NIKOLJUK, V. F. (1969). Some aspects of the study of soil Protozoa. *Acta Protozool.*, **7**, 99–109.

188. NOIROT-TIMOTHÉE, C. (1959). Recherches sur l'ultrastructure d'*Opalina ranarum*. *Ann. Sci. Nat., Zool.*, 12ᵉ serie, **1**, 265–281.

189. NOLAND, L. E. (1959). Ciliophora, pp. 265–297. In *Fresh-Water Biology*, EDMONDSON, W. T., Wiley, New York.

190. NOLAND, L. E. and GOJDICS, M. (1967). Ecology of free-living Protozoa, pp. 215–266. In *Research in Protozoology*, **2**, CHEN, T.-T., Pergamon Press, Oxford.

191. NORTHCOTE, D. H. (1971). The Golgi apparatus. *Endeavour*, **30**, 26–33.

192. OKAJIMA, A. (1953). Studies on the metachronal wave in *Opalina* I. Electrical stimulation with the micro-electrode. *Jap. J. Zool.*, **11**, 87–100.

192a. PAL, R. A. (1972). The osmoregulatory system of the amoeba *Acanthamoeba castellanii*. *J. exp. Biol.*, **57**, 55–76.

193. PARDUCZ, B. (1967). Ciliary movement and coordination in Ciliates. *Int. Rev. Cytol.*, **21**, 91–128.

194. PARKE, M. and ADAMS, I. (1960). The motile (*Crystallolithus hyalinus* Gaardner and Markali) and non-motile phases in the life history of *Coccolithus pelagicus* (Wallich) Schiller. *J. mar. biol. Ass. U.K.*, **39**, 263–274.

195. PARKE, M., MANTON, I. and CLARKE, B. (1955). Studies on marine flagellates II. Three new species of *Chrysochromulina*. *J. mar. biol. Ass. U.K.*, **34**, 579–609.

196. PICKEN, L. E. R. (1937). The structure of some protozoan communities. *J. Ecol.*, **25**, 368–384.

197. PITELKA, D. R. (1961). Observations on the kinetoplast-mitochondrion and the cytostome of *Bodo*. *Exp. Cell Res.*, **25**, 87–93.

198. PITELKA, D. R. (1963). *Electron-microscopic Structure of Protozoa*. Pergamon Press, Oxford.

199. PITELKA, D. R. (1965). New observations on cortical ultrastructure in *Paramecium*. *J. Microscopie*, **4**, 373–394.

200. PITELKA, D. R. (1969). Fibrillar systems in Protozoa, pp. 280–388. In *Research in Protozoology*, **3**, CHEN, T.-T., Pergamon Press, Oxford.

201. POISSON, R. (1953). Sous-embranchement des Cnidosporidies, pp. 1006–1088. In *Traité de Zoologie*, **1**(ii), GRASSÉ, P.-P., Masson, Paris.

202. POLLARD, T. D. and ITO, S. (1970). Cytoplasmic filaments of *Amoeba proteus*. I. The role of filaments in consistency change and movement. *J. Cell Biol.*, **46**, 267–289.

203. POLLARD, T. D., SHELTON, E., WEIHING, R. R. and KORN, E. D. Ultrastructural characterization of F-actin isolated from *Acanthamoeba castellanii* and identification of cytoplasmic filaments as F-actin by reaction with rabbit heavy meromyosin. (1970). *J. molec. Biol.*, **50**, 91–97.

204. PREER, J. R. (1969). Genetics of the Protozoa, pp. 129–278. In *Research in Protozoology*, **3**, CHEN, T.-T., Pergamon Press, Oxford.

205. PRESCOTT, D. M. and STONE, G. E. (1967). Replication and function of the protozoan nucleus, pp. 117–146, In *Research in Protozoology*, **2**, CHEN, T.-T., Pergamon Press, Oxford.

206. de PUYTORAC, P. (1954). Contribution à l'étude cytologique et taxonomique des infusoires astomes. *Ann. Sci. nat. Zool.*, 11e serie, **16**, 85–270.

207. de PUYTORAC, P. (1959). Le cytosquellette et les systèmes fibrillaires du cilié *Metaradiophrya gigas* de Puytorac d'après étude au microscope électronique. *Arch. Anat. micr. Morph. exp.*, **48**, 49–62.

208. RAABE, Z. (1970). Ordo Thigmotricha (Ciliata-Holotricha) III. Familiae Ancistrocomidae et Sphenophryidae. *Acta Protozool*, **7**, 385–463.

209. RAIKOV, I. B. (1969). The macronucleus of ciliates, pp. 1–128. In *Research in Protozoology*, **3**, CHEN, T.-T., Pergamon Press, Oxford.

210. RANDALL, J. T. and JACKSON, S. F. (1958). Fine structure and function in *Stentor polymorphus*. *J. biophys. biochem. Cytol.*, **4**, 807–830.

211. RIKMENSPOEL, R. and SLEIGH, M. A. (1970). Bending movements and elastic constants in cilia. *J. Theor. Biol.*, **28**, 81–100.

212. RINGO, D. L. (1967). The arrangement of subunits in flagellar fibres. *J. Ultrastruct. Res.*, **17**, 266–277.

213. ROBERTSON, J. D. (1969). Molecular structure of biological membranes, pp. 1403–1443. In *Handbook of Molecular Cytology*. LIMA-DE-FARIA, A., North Holland, Amsterdam.

214. ROLLO, I. M. (1964). The chemotherapy of malaria, pp. 525–561. In *Biochemistry and Physiology of Protozoa*, 3, HUTNER, S. H., Academic Press, New York.

215. ROTH, L. E., PIHLAJA, D. J. and SHIGENAKA, Y. (1970). Microtubules in the heliozoan axopodium. 1. The gradion hypothesis of allosterism in structural proteins. *J. Ultrastruct. Res.*, 30, 7–37.

216. ROUND, F. E. (1973). *The Biology of the Algae*. 2nd Edn. Edward Arnold, London.

217. RUDZINSKA, M. A. (1965). The fine structure and function of the tentacle in *Tokophrya infusionum. J. Cell Biol.*, 25, 459–477.

218. RUDZINSKA, M. A. (1969). The fine structure of malaria parasites. *Int. Rev. Cytol.*, 25, 161–199.

219. RUDZINSKA, M. A. and TRAGER, W. (1962). Intracellular phagotrophy in *Babesia rodhaini* as revealed by electron microscopy. *J. Protozool.*, 9, 279–288.

220. RUDZINSKA, M. A. and VICKERMAN, K. (1968). The fine structure, pp. 217–306. In *Infectious Blood Diseases of Man and Animals*, 1, WEINMAN, D. and RISTIC, M. Academic Press, New York.

221. SANDON, H. (1927). *The Composition and Distribution of the Protozoan Fauna of the Soil*. Oliver and Boyd, Edinburgh.

222. SANDON, H. (1963). *Essays on Protozoology*. Hutchinson, London.

223. SATIR, P. (1968). Studies on cilia. III. Further studies on the cilium tip and a 'sliding filament' model of ciliary motility. *J. Cell Biol.*, 39, 77–94.

233a. SATIR, B., SCHOOLEY, C. and SATIR, P. (1973). Membrane fusion in a model system. Mucocyst secretion in *Tetrahymena J. Cell Biol.*, 56, 153–176.

224. SCHERBAUM, O. H. and LOEFER, J. B. (1964). Environmentally induced growth oscillations in Protozoa, pp. 9–59. In *Biochemistry and Physiology of Protozoa*, 3, HUTNER, S. H., Academic Press, New York.

225. SCHMIDT-NIELSEN, B. and SCHRAUGER, C. R. (1963). *Amoeba proteus*: studying the contractile vacuole by micropuncture. *Science, N.Y.*, 139, 606–607.

226. SCHNEIDER, L. (1960). Elektronenmikroskopische Untersuchungen über der Nephridialsystem von *Paramecium. J. Protozool.*, 7, 75–101.

227. SCHREVEL, J. (1970). Contribution à l'étude des *Selenidiidae* parasites d'annélides polychètes. I. Cycles biologiques. *Protistologica*, 6, 389–426.

228. SCHREVEL, J. (1971). Contribution à l'étude des *Selenidiidae* parasites d'annélides polychètes. II. Ultrastructure de quelques trophozoites. *Protistologica*, 7, 101–130.

229. SCHWAB, D. (1969). Elektronenmikroskopische Untersuchung an der Foraminifere *Myxotheca arenilega* Schaudinn. *Z. Zellforsch.*, 96, 295–324.

230. SERAVIN, L. N. (1971). Mechanisms and coordination of cellular locomotion. *Comp. Physiol. Bioch.*, 4, 37–111.

231. SIEGEL, R. W. (1967). Genetics of ageing and the life cycles in ciliates. *Symp. Soc. exp. Biol.*, 21, 127–148.

232. SLEIGH, M. A. (1962). *The Biology of Cilia and Flagella*. Pergamon Press, Oxford.

233. SLEIGH, M. A. (1964). Flagellar movement of the sessile flagellates *Actinomonas, Codonosiga, Monas*, and *Poteriodendron. Q. Jl microsc. Sci.*, **105**, 405–414.

234. SLEIGH, M. A. (1966). The coordination and control of cilia. *Symp. Soc. exp. Biol.*, **20**, 11–31.

235. SLEIGH, M. A. (1968). Patterns of ciliary beating. *Symp. Soc. exp. Biol.*, **22**, 131–150.

236. SLEIGH, M. A. (1970). Some factors affecting the excitation of contraction in *Spirostomum. Acta Protozool.*, **7**, 335–352.

237. SLEIGH, M. A. (1971). Cilia. *Endeavour*, **30**, 11–17.

238. SLEIGH, M. A. (1973). Cell Motility, pp. 525–569. In *Cell Biology in Medicine*, BITTAR, E. E., Wiley, New York.

239. SLEIGH, M. A. (1973). *Cilia and Flagella*, Academic Press, London. In press.

240. SLEIGH, M. A. and HOLWILL, M. E. J. (1969). Energetics of ciliary movement in *Sabellaria* and *Mytilus. J. exp. Biol.*, **50**, 733–743.

241. SMALL, E. B. (1967). The Scuticociliatida, a new order of the class Ciliatea (phylum Protozoa, subphylum Ciliophora). *Trans. Am. microsc. Soc.*, **86**, 345–370.

242. SMALL, E. B. and MARSZALEK, D. S. (1969). Scanning electron microscopy of fixed, frozen, and dried Protozoa. *Science, N.Y.*, **163**, 1064–1065.

243. SMITH, D. C., MUSCATINE, L. and LEWIS, D. (1969). Carbohydrate movement from autotrophs to heterotrophs in parasitic and mutualistic symbiosis. *Biol. Rev.*, **44**, 17–90.

244. SONNEBORN, T. M. (1957). Breeding systems, reproductive methods and species problems in Protozoa, pp. 155–324. In *The Species Problem*, MAYR, E., Am. Ass. Adv. Sci., Washington.

245. SPRAGUE, V. and VERNICK, S. H. (1968). Light and electron microscope study of a new species of *Glugea* (Microsporida, Nosematidae) in the 4-spined stickleback *Apeltes quadracus. J. Protozool.*, **15**, 547–571.

246. STEPHENS, R. E. (1970). Thermal fractionation of outer fiber doublet microtubules into A- and B-subfiber components: A and B tubulin. *J. molec. Biol.*, **47**, 353–363.

247. STEVENS, B. J. and ANDRÉ, J. (1969). The nuclear envelope, pp. 837–871. In *Handbook of Molecular Cytology*, LIMA-DE-FARIA, A., North Holland, Amsterdam.

248. STOCKEM, W. and WOHLFARTH-BOTTERMANN, K. E. (1969). Pinocytosis (Endocytosis), pp. 1373–1409. In *Handbook of Molecular Cytology*, LIMA-DE-FARIA, A., North Holland, Amsterdam.

249. STOCKEM, W. and WOHLFARTH-BOTTERMANN, K. E. (1970). On the fine structure of the trichocysts of *Paramecium. Cytobiol.*, **1**, 420–436.

250. STOUT, J. D. and HEAL, O. W. (1967). Protozoa, pp. 149–195. In *Soil Biology*, BURGES, N. A. and RAW, F., Academic Press, London.

251. SWALE, E. M. F. (1969). A study of the nannoplankton flagellate *Pedinella hexacostata* Vysotskii by light and electron microscopy. *Br. phycol. J.*, **4**, 65–86.

252. TALIAFERRO, W. H. and STAUBER, L. A. (1969). Immunology of protozoan infections, pp. 505–564. In *Research in Protozoology*, **3**, CHEN, T.-T., Pergamon Press, Oxford.

253. TAMM, S. L. and HORRIDGE, G. A. (1970). The relation between the orientation of the central fibrils and the direction of beat in cilia of *Opalina. Proc. R. Soc. Lond. B.*, **175**, 219–233.

254. TARTAR, V. (1961). *The Biology of Stentor*. Pergamon Press, Oxford.

255. TARTAR, V. (1966). Stentors in dilemmas. *Zeit. allg. Mikrobiol.*, **6**, 125–134.
256. TARTAR, V. (1967). Morphogenesis in Protozoa, pp. 1–116. In *Research in Protozoology*, **2**, CHEN, T.-T., Pergamon Press, Oxford.
257. TARTAR, V. (1968). Micrurgical experiments on cytokinesis in Stentor *coeruleus. J. exp. Zool.*, **167**, 21–36.
258. TAYLOR, D. L. (1969). The nutritional relationship of *Anemonia sulcata* (Pennant) and its Dinoflagellate symbiont. *J. Cell Sci.*, **4**, 751–762.
259. THOMPSON, R. H. (1959). Algae, pp. 115–170. In *Fresh-Water Biology*. EDMONDSON, W. T., Wiley, New York.
260. TILNEY, L. G. and BYERS, B. (1969). Studies on the microtubules in Heliozoa. V. Factors controlling the organization of microtubules in the axonomal pattern in *Echinosphaerium* (*Actinosphaerium*) *nucleofilum J. Cell Biol.*, **43**, 148–165.
261. TRÉGOUBOFF, G. (1953). Actinopodes, pp. 267–489. In *Traité de Zoologie*, **1**(ii), GRASSÉ, P.-P., Masson, Paris.
262. TUCKER, J. B. (1967). Changes in nuclear structure during binary fission in the ciliate *Nassula. J. Cell Sci.*, **2**, 481–498.
263. TUCKER, J. B. (1968). Fine structure and function of the cytopharyngeal basket in the ciliate *Nassula. J. Cell Sci.*, **3**, 493–514.
264. TUCKER, J. B. (1970). Morphogenesis of a large microtubular organelle and its association with basal bodies in the ciliate *Nassula. J. Cell Sci.*, **6**, 385–429.
265. TUCKER, J. B. (1971). Microtubules and a contractile ring of microfilaments associated with a cleavage furrow. *J. Cell Sci.*, **8**, 557–571.
266. TUFFRAU, M. (1964). Quelques variantes techniques de l'imprégnation des Ciliés par le protéinate d'argent. *Arch. Zool. Exp. Gen.*, **104**, 186–190.
267. TUFFRAU, M. (1967). Les structures fibrillaires somatiques et buccales chez les ciliés Hétérotriches. *Protistologica*, **3**, 369–394.
268. TUFFRAU, M. (1970). Nouvelles observations sur l'origine du primordium buccal chez les hypotriches. *C.r. hebd. Séanc. Acad. Sci.*, Paris, **270**, 104–107.
269. TUFFRAU, M., PYNE, C. K. and de HALLER, G. (1968). Organisation de l'infraciliature chez quelques ciliés hypotriches. *Protistologica*, **4**, 289–301.
270. VALKENBURG, S. D. van (1971). Observations on the fine structure of *Dictyocha fibula* Ehrenberg. I. The skeleton. II. The protoplast. *J. Phycol.*, **7**, 113–132.
271. VICKERMAN, K. (1962). The mechanism of cyclical development in trypanosomes of the *Trypanosoma brucei* sub-group: An hypothesis based on ultrastructural observations. *Trans. R. Soc. trop. Med. Hyg.*, **56**, 487–495.
272. VICKERMAN, K. (1969). The fine structure of *Trypanosoma congolense* in its bloodstream phase. *J. Protozool.*, **16**, 54–69.
273. VICKERMAN, K. (1969). On the surface coat and flagellar adhesion in Trypanosomes. *J. Cell Sci.*, **5**, 163–194.
274. VICKERMAN, K. and COX, F. E. G. (1967). *The Protozoa*. John Murray, London.
275. VICKERMAN, K. and PRESTON, T. M. (1970). Spindle microtubules in the dividing nuclei of Trypanosomes. *J. Cell Sci.*, **6**, 365–383.
276. VINCKIER, D., DEVAUCHELLE, G. and PRENSIER, G. (1970). *Nosema vivieri* n. sp. (Microsporida, Nosematidae) hyperparasite d'une Grégarine

vivant dans le coelome d'une Némerte. *C.r. hebd. Seanc. Acad. Sci.,* Paris, **270**, 821–823.
277. VIVIER, E. (1969). L'ultrastructure des formes vegetatives des sporozoaires. *Proc. 3rd Int. Congr. Protozool.,* Leningrad, 47–49.
278. VIVIER, E., DEVAUCHELLE, G., PETITPREZ, A., PORCHET-HENNERE, E., PRENSIER, G., SCHREVEL, J. and VINCKIER, D. (1970). Observations de cytologie comparée chez les sporozoaires. *Protistologica,* **6**, 127–150.
279. VIVIER, E. and PETITPREZ, A. (1969). Observations ultrastructurales sur l'hematozoaire *Anthemosoma garnhami* et examen de critères morphologiques utilisables pour la taxonomie chez les sporozoaires. *Protistologica,* **5**, 363–379.
280. WAGTENDONK, W. J. van. (1955). Encystment and excystment of Protozoa, pp. 85–90. In *Biochemistry and Physiology of Protozoa,* **2**, HUTNER, S. H. and LWOFF, A., Academic Press, New York.
281. WARNER, F. (1970). New observations on flagellar fine structure. The relationship between matrix structure and the microtubule component of the axoneme. *J. Cell Biol.,* **47**, 159–182.
282. WARR, J. R., MCVITTIE, A., RANDALL, J. T. and HOPKINS, J. M. (1966). Genetic control of flagellar structure in *Chlamydomonas reinhardii. Genet. Res., Camb.,* **7**, 335–351.
283. WATTERS, C. (1968). Studies on the motility of the Heliozoa. I. The locomotion of *Actinosphaerium eichhorni* and *Actinophrys* sp. *J. Cell Sci.,* **3**, 231–244.
284. WATTIAUX, R. (1969). Biochemistry and function of lysosomes, pp. 1159–1178. In *Handbook of Molecular Cytology.* LIMA-DE-FARIA, A., North Holland, Amsterdam.
285. WEBB, M. (1956). An ecological study of brackish water ciliates. *J. Anim. Ecol.,* **25**, 148–175.
286. WESSENBERG, H. (1961). Studies on the life cycle and morphogenesis of *Opalina. Univ. Calif. Publ. Zool.,* **61**, 315–370.
287. WESSENBERG, H. (1966). Observations on cortical ultrastructure in *Opalina. J. Microscopie,* **5**, 471–492.
288. WESSENBERG, H. and ANTIPA, G. (1968). Studies on *Didinium nasutum.* 1. Structure and ultrastructure. *Protistologia,* **4**, 427–447.
289. WESSENBERG, H. and ANTIPA, G. (1970). Capture and ingestion of *Paramecium* by *Didinium nasutum. J. Protozool.,* **17**, 250–270.
290. WOHLMAN, A. and ALLEN, R. D. (1968). Structural organization associated with pseudopod extension and contraction during cell locomotion in *Difflugia. J. Cell Sci.,* **3**, 105–114.
291. WOLKEN, J. J. (1961). *Euglena.* Rutgers Univ. Press, New Brunswick, N.J.
292. ZEUTHEN, E. and RASMUSSEN, L. (1972). Synchronized cell division in Protozoa, pp. 9–145. In *Research in Protozoology,* **4**, CHEN, T.-T., Pergamon Press, Oxford.
293. ZEUTHEN, E. and SCHERBAUM, O. (1954). Synchronous divisions in mass cultures of the ciliate protozoan *Tetrahymena pyriformis,* as induced by temperature changes. *Colston pap.,* **7**, 141–157.
294. CORLISS, J. O. (1974). The changing world of ciliate systematics : Historical analysis of past efforts and a newly proposed phylogenetic scheme of classification for the protistan phylum Ciliophora. *Syst. Zool.,* **23**, 91–138.

Index